# R Companion for

# Sampling

## Design and Analysis
### Third Edition

# *R Companion for*

# Sampling

## Design and Analysis
### Third Edition

Yan Lu and Sharon L. Lohr

CRC Press
Taylor & Francis Group
Boca Raton London New York

CRC Press is an imprint of the
Taylor & Francis Group, an **informa** business

A CHAPMAN & HALL BOOK

First edition published 2022
by CRC Press
6000 Broken Sound Parkway NW, Suite 300, Boca Raton, FL 33487-2742

and by CRC Press
2 Park Square, Milton Park, Abingdon, Oxon, OX14 4RN

**Library of Congress Cataloging-in-Publication Data**

Names: Lu, Yan, author. | Lohr, Sharon L., 1960- author.
Title: R companion for sampling : design and analysis / Yan Lu and Sharon L. Lohr.
Description: First edition. | Boca Raton : CRC Press, 2022. | Includes bibliographical references and index. | Summary: "The R Companion for Sampling: Design and Analysis, designed to be read alongside Sampling: Design and Analysis, Third Edition by Sharon L. Lohr (SDA; 2022, CRC Press), shows how to use functions in base R and contributed packages to perform calculations for the examples in SDA. No prior experience with R is needed. Chapter 1 tells you how to obtain R and RStudio, introduces basic features of the R statistical software environment, and helps you get started with analyzing data. Each subsequent chapter provides step-by-step guidance for working through the data examples in the corresponding chapter of SDA, with code, output, and interpretation. Tips and warnings help you develop good programming practices and avoid common survey data analysis errors. R features and functions are introduced as they are needed so you can see how each type of sample is selected and analyzed. Each chapter builds on the knowledge developed earlier for simpler designs; after finishing the book, you will know how to use R to select and analyze almost any type of probability sample"-- Provided by publisher.
Identifiers: LCCN 2021039318 (print) | LCCN 2021039319 (ebook) | ISBN 9781032135946 (paperback) | ISBN 9781032132150 (hardback) | ISBN 9781003228196 (ebook)
Subjects: LCSH: R (Computer program language) | Sampling (Statistics)
Classification: LCC QA276.45.R3 L8 2022 (print) | LCC QA276.45.R3 (ebook) | DDC 519.5/202855133--dc23
LC record available at https://lccn.loc.gov/2021039318
LC ebook record available at https://lccn.loc.gov/2021039319

ISBN: 978-1-032-13215-0 (hbk)
ISBN: 978-1-032-13594-6 (pbk)
ISBN: 978-1-003-22819-6 (ebk)

DOI: 10.1201/9781003228196

Access the Support Material: https://www.routledge.com/9781032135946

*To Guoyi and Lynn, and to Doug*

# Contents

# Preface

*R Companion for Sampling: Design and Analysis, Third Edition* shows how to use the R statistical software environment to perform the calculations in the textbook *Sampling: Design and Analysis, Third Edition* (SDA) by Sharon L. Lohr. It is intended to be read in conjunction with SDA and is not a standalone text. The parallel book by Lohr (2022) shows how to perform the computations for the examples using SAS® software, and could be read together with this book and SDA to learn how to perform the analyses in each software package.

All code and data sets can be downloaded from any of the following websites:

> `https://math.unm.edu/~luyan/rbook.html`
>
> `https://www.sharonlohr.com`
>
> `https://www.routledge.com/9781032135946`

The first two websites also contain additional programs, not discussed in this book, that you can adapt for some of the SDA exercises. The data sets used in this book have also been saved in R format in the contributed R package `SDAResources` (Lu and Lohr, 2021).

In this book, we give step-by-step guidance for using functions from base R and contributed packages to select samples and analyze the data sets discussed in Chapters 1–13 of SDA. The software, however, can do much more than analyze the examples presented in this book. You can find information on advanced capabilities for the `survey` and `sampling` contributed packages in the documentation for those packages by Lumley (2020) and Tillé and Matei (2021); the books and articles by Lumley (2004, 2010) and Tillé and Matei (2010) provide additional information about the packages. Goga (2018) gives an overview of using R for survey sampling.

For easy reference, the index at the back of the book gives page numbers for the examples in SDA. To locate the code and output for Example 2.5, for example, look up the subentry "Example 02.05" under "Examples in SDA" in the index. The book also gives code and suggestions for some of the exercises in SDA, and these are listed under index entry "Exercises in SDA."

Each chapter ends with a summary section containing tips and warnings for the analyses discussed in that chapter. These provide ways of avoiding common survey data analysis errors and checking whether you did the analysis correctly.

Although prior experience with R is helpful, it is not needed to read this book. Chapter 1 tells how to obtain the software and do basic operations in R. It also lists resources for learning more about programming in R and tells how to obtain help.

This book makes use of functions that exist in base R and contributed packages, and does not discuss how to write R functions. One of R's most valuable features, however, is the capacity for writing functions to carry out new tasks. Advanced R users may want to write their own functions to select samples or analyze data from a complex survey. When teaching survey sampling to students who have R programming experience, we have sometimes asked them to write their own functions to carry out various sampling tasks. This helps solidify their knowledge of the material and allows them to do computations not available in existing functions. For example, we have asked students to write R functions to perform allocation for and analyze data from a stratified random sample, select a with-replacement unequal-probability sample using Lahiri's method, compute the Sen–Yates–Grundy estimate of the variance, simulate the sampling distribution of a statistic, and find empirical estimates of the coverage probability of a confidence interval for a biased estimator.

All code, data sets, and output in this book are provided for educational purposes only and without warranty. Base R does not contain functions for survey data, and this book relies heavily on contributed packages that have been developed. These packages are in widespread use and have been quality-checked by their authors and other users. We have verified that the calculations from the R functions used for the examples in this book agree with calculations by the formulas and with calculations performed in other survey software packages.

Other R packages may not be checked as carefully, however. Although R contributed packages undergo some consistency and functionality tests when they are submitted (see Wickham, 2015, for a description of checks that are performed), no central authority reviews the packages to make sure that the functions do what they claim or that the algorithms perform computations accurately. Most R contributed packages are not peer-reviewed, and you should be aware that some may contain errors.

The code and output in this book were developed using version 4.0.4 of R for Windows (R Core Team, 2021) and the versions of the packages listed in their respective bibliography entries, and all code in the book works with those versions. But R is a dynamic language, and the R Core Team and authors of contributed packages can change or remove functions at any time. Although most authors who revise a package try to avoid changes that will affect previously written code, functions in R are not guaranteed to be backward compatible—it is possible that R code you write today may not work the same way with future versions of the software. If backward compatibility is important to you—for example, if you will be using the same code to produce estimates each year for an annual survey—you may want to perform or check your computations in a package that is backward compatible, such as SAS software. If a function changes in a subsequent version of an R package, you can either:

- Read the documentation and change your code so that it works with the modified function, or

- Download and use the older version of the package. You can find previous versions on the package's web page under the heading "Old sources."

**Acknowledgments.** Many thanks to John Kimmel, our editor at CRC Press, for encouraging us to write this book, and to the CRC Press production team for all their support and help. We are grateful to Yves Tillé and Thomas Lumley for answering questions about the `sampling` and `survey` packages. Students in Yan Lu's sampling class at the University of New Mexico provided helpful suggestions for clarifying the material. We also want to thank Lynn Zhang for helping with the preparation of the `SDAResources` package.

# 1

## Getting Started

The R statistical software environment is a powerful and flexible platform for performing statistical analyses. The basic package contains thousands of functions for computing statistics, and user-contributed packages for this open-source software provide thousands more. Advanced users can write their own functions to implement new methods for statistical analyses.

Best of all, the base R package and all user-contributed packages are available free of charge to anyone with an internet connection.

This chapter tells you how to obtain R software and contributed packages and introduces you to some basic R functions. It also shows you how to read data sets into R and save output and graphics produced while you are using the package.

**Conventions used in this book.** This book is intended to be read in conjunction with *Sampling: Design and Analysis, Third Edition* by Sharon L. Lohr, henceforth referred to as SDA. Many of the examples in this book refer to figures, tables, examples, or exercises in SDA. To avoid confusion, we refer to figures in SDA as "Figure x.x in SDA." We refer to figures in *this* book as "Figure x.x" with no qualifier.

The names of external data files and programs, such as `agsrs.csv` and `ch02.R`, are in `typewriter font`, as are the names of R packages and code we type. Variable names, function names, and internal R data set names are in *italic type.*

Much of this book consists of R commands and output, set in light shaded boxes such as the following:

```
# This is a comment
# Enter data values into vector 'myvec'
myvec <- c(1,2,3,9,14,27,5,21,pi)
# Print the vector 'myvec'
myvec
## [1]  1.000000  2.000000  3.000000  9.000000 14.000000 27.000000  5.000000
## [8] 21.000000  3.141593
# Calculate summary statistics
summary(myvec)
##    Min. 1st Qu.  Median    Mean 3rd Qu.    Max.
##    1.00    3.00    5.00    9.46   14.00   27.00
```

To distinguish between the code and the output that is produced, a command that we typed into R is displayed flush against the left margin. The output from that command is preceded by `##`. Our comments are preceded by a single `#`. You can obtain the commands and comments (without the output) for all code in files `ch01.R`, `ch02.R`, etc., on the book website (see the Preface for the website address).

DOI: 10.1201/9781003228196-1

## 1.1   Obtaining the Software

The R system is available for FREE download from the Comprehensive R Archive Network at

https://cran.r-project.org/.

If R is not already installed on your system, download the package now. Click on the link for the operating system of your choice: Windows, Mac, or Linux. Install the package (you can do this for Windows or Mac by double-clicking on the icon after the download is finished). You should be able to click "Next" for all dialogs to finish the installation.

Good! Now you can use R directly or through the integrated development environment RStudio®. RStudio adds a larger number of statistical packages to R and provides a user-friendly graphical interface. You can obtain RStudio free of charge for Windows, Mac, or Linux from https://rstudio.com. As with R, you can install RStudio by double-clicking on the icon after download and can click on "Next" for all dialogs. If you have any trouble, numerous videos are available online that demonstrate the installation of R and RStudio.

Figure 1.1 is a screenshot of RStudio. It shows four windows:

1. Upper-left: Program editor window, where you will write code and programs. You can save the contents of this window so you have a record of what you have done.
2. Lower-left: R console, where you will run R commands.
3. Upper-right: The Workspace window tells you what you have created so far. Environment lists the variables in memory. History lists the commands you have submitted in the R console. The Tutorial window provides a tutorial for using R and RStudio.
4. Lower-right: Displays the files in the current working directory, plots you have produced, packages installed and whether they are loaded, and help for commands.

**How to get things done in RStudio**

1. Work out commands in the program editor window, and then either copy/paste the commands into the console or use Ctrl-Enter or click the icon "Run" to submit the current line or selection. You can open a new program editor window by clicking "R Script" under the "New File" icon in the upper left of the main RStudio toolbar (this can also be accessed through the File menu).
2. The program editor will keep a history of what you have done.
3. Make comments using # to help your future self recall what you did and why you did it.

**Updating R and RStudio.**   If you already have R and RStudio installed, you may update them to the latest versions. You can either reinstall the latest version or (only for Windows) use the *updateR* function from contributed package installr. To update RStudio, select "Check for Updates" from the Help menu within the program.

## 1.2   Installing R Packages

The base R software contains few functions for analyzing survey data. In this book we rely on user-contributed packages that provide functions for selecting probability samples and for graphing and analyzing survey data.

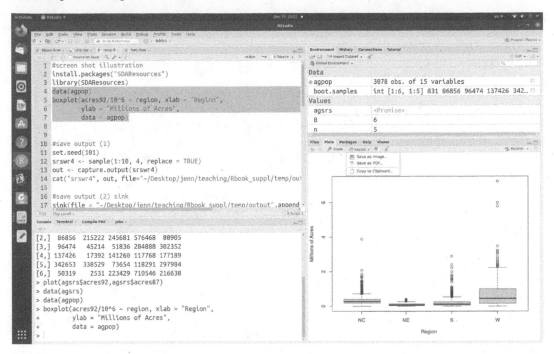

**FIGURE 1.1:** Screenshot of RStudio.

The main contributed packages used in this book are:

- `survey` (Lumley, 2020): Used to calculate statistics from complex surveys with functions such as *svydesign*, *svymean*, and *svytotal*.

- `sampling` (Tillé and Matei, 2021): Used to select probability samples. The `sampling` package will also analyze data from complex samples, although in this book we primarily use the `survey` package for data analysis.

- `SDAResources` (Lu and Lohr, 2021): Includes all the data sets of SDA in R format and additional functions for analyzing and graphing probability samples.

Install these packages now. You only need to install each package once for a particular version of R (you may need to reinstall packages if you update to a new R version).

Copy and paste the following code into the R Console window and press [Enter]. Each package will be downloaded (along with package dependencies) and installed.

```
install.packages("survey")
install.packages("sampling")
install.packages("SDAResources")
```

You must load a package into the program with the *library* function each time you open a new R session. When they are needed, run (in this order):

```
library(survey)
library(sampling)
library(SDAResources)
```

## 1.3 R Basics

**Getting help.** There are many sources for help about R and RStudio. The web pages
https://www.r-project.org/ and https://education.rstudio.com/ contain links to
online tutorials and resources. For those who prefer a printed book, Horton and Kleinman
(2015), Horton et al. (2018), Ismay and Kennedy (2019), and Kabacoff (2021) provide useful
introductions to getting started with R and RStudio. The book by Wickham (2019) is an
advanced guide to R.

The *help* function in R provides access to the documentation pages for R functions, data
sets, and other objects. For example, if you want to get help with the function *mean*, type
`help("mean")`, or type `help.search("mean")` into the R console to search the help system
for documentation matching the given character string "mean". You can also display the
code of a function (for some packages) by typing its name. For example, typing `ls` in the R
console window displays the code for the function *ls*, which lists the objects in the directory;
to list the objects, type `ls()`.

For another example, suppose that you have not installed package `survey` yet, and
you want to get help with the function *svymean*. Typing `help("svymean")` will show
that "No documentation for 'svymean' in specified packages and libraries". Typing
`help.search("svymean")` will direct you to the page with information that function
*svymean* is under package `survey`.

To find help for a package, for example, package `survey`, type `help(package = "survey")`.
To find help for data *agpop*, type `help("agpop")`.

In RStudio, you can find help on a function by typing the function name into the search
window of the Help menu.

**Frequently used commands.** Tables 1.1 and 1.2 list frequently used operators, functions,
and commands on vectors, matrices, and data frames. Note that R is case-sensitive: *myvar*,
*MYVAR*, and *Myvar* are all different variables.

The R functions we use to analyze survey data operate on data frames. A data frame is
like a matrix, where each row contains the data values for an observation, and each column
contains the data values for one variable. You can turn a matrix $A$ into a data frame with
the *as.data.frame* function.

```
A<-matrix(0,2,2)  # create a 2*2 matrix with all 0s
A[1,1]<-2
A[1,2]<-5
A[2,1]<-1
A[2,2]<-4
colnames(A)<-c("y","x")
A
##     y x
## [1,] 2 5
## [2,] 1 4
A.data <- as.data.frame(A) # turn matrix A into a data frame A.data
A.data
##   y x
## 1 2 5
## 2 1 4
```

**TABLE 1.1**
Frequently used commands in R: Vectors, matrices, and mathematical functions.

| Command | Purpose | Example | Output |
|---|---|---|---|
| **Assignment and Comment** | | | |
| <- | Assign value | a <- 1 | 1 |
| = | Assign value | b = 1 + 2 | 3 |
| # | Comment | # My comment | |
| **Mathematical Functions** | | | |
| + | Addition | 1 + 1 | 2 |
| − | Subtraction | 1 − 1 | 0 |
| * | Scalar multiplication | 1*2 | 2 |
| / | Division | 1/2 | 0.5 |
| ^ | Exponentiation | $2^3$ | 8 |
| abs | Absolute value | abs(−3) | 3 |
| exp | Exponential function | exp(1) | 2.718282 |
| log | Natural log | log(2.718282) | 1 |
| sqrt | Square root | sqrt(4) | 2 |
| **Vector/Matrix Operations** | | | |
| c | Combine values into vector | vec1<-c(1,2,3) vec2<-c(4,5,6) | 1 2 3 <br> 4 5 6 |
| seq | Sequence or, when by=1 | seq(from=1,to=10,by=2) 1:5 | 1 3 5 7 9 <br> 1 2 3 4 5 |
| sequence | Vector of sequences | sequence(vec1) | 1 1 2 1 2 3 |
| rep | Replicate | rep(1,4) vec3<-rep(vec1,2) | 1 1 1 1 <br> 1 2 3 1 2 3 |
| cbind | Column bind | mat1<-cbind(vec1,vec2) | 1 4 <br> 2 5 <br> 3 6 |
| rbind | Row bind | mat2<-rbind(vec1,vec2) | 1 2 3 <br> 4 5 6 |
| length | Length of vector | length(vec1) | 3 |
| sort | Sort a vector | sort(vec3) | 1 1 2 2 3 3 |
| order | Indices to sort a vector | order(vec3) | 1 4 2 5 3 6 |
| unique | List unique objects | unique(vec3) | 1 2 3 |
| sum | Summation | sum(c(1,2,4)) | 7 |
| prod | Product | prod(c(1,2,4)) | 8 |
| %*% | Matrix multiplication | mat1 %*% mat2 | 17 22 27 <br> 22 29 36 <br> 27 36 45 |
| t | Transpose | t(mat1) | 1 2 3 <br> 4 5 6 |

## 1.4 Reading Data into R

The first step for using R to analyze data from a survey is to read the data into the system. There are four basic ways to do this.

**Enter the data directly into R.** You can use the *c* function to enter a vector directly from the R console. Type

```
myvec <- c(5,2,8,4)
```

**TABLE 1.2**
Frequently used commands in R: Extracting data elements and computing statistics.

| Command | Purpose | Example | Output |
|---|---|---|---|
| **Working with Data** | | | |
| as.data.frame | Create data frame | A<-as.data.frame(mat1) | vec1  vec2<br>1      4<br>2      5<br>3      6 |
| names(data) | Extract variable names | names(A) | "vec1" "vec2" |
| $ | Extract a column by name | A$vec1 | 1 2 3 |
| [ , j] | Extract a column by number | A[,1] | 1 2 3 |
| [ i , ] | Extract a row by number | A[1,] | 1 4 |
| nrow(data) | # of rows in *data* | nrow(A) | 3 |
| ncol(data) | # of columns in *data* | ncol(A) | 2 |
| head(data,n=i) | First i rows | head(A,n=2) | 1  4<br>2  5 |
| tail(data, n=i) | Last i rows | tail(A,n=2) | 2  5<br>3  6 |
| **Statistics** | | | |
| max/min | Maximum/Minimum of<br>a vector or matrix | max(vec1)<br>min(mat1) | 3<br>1 |
| quantile | Calculate quantiles | quantile(1:101,probs=c(0,0.25,0.5)) | 0%  25%  50%<br>1    26   51 |
| mean | Sample mean | mean(vec1) | 2 |
| var | Sample variance | var(vec1) | 1 |
| cor | Correlation of two vectors | cor(vec1,vec2) | 1 |

to store the vector of values (5, 2, 8, 4) in variable *myvec*. You can also use the *data.frame* function to enter data into a data frame from the R console:

```
# read matrix into data frame "mydata"
mydata <- data.frame(x = c(1,4,3,2), y = c(2,4,2,4), z = c(3,7,2,3))
# show the data
mydata
##   x y z
## 1 1 2 3
## 2 4 4 7
## 3 3 2 2
## 4 2 4 3
```

**Read data from a comma-delimited (.csv) or text file.** With a longer data set, it is often more convenient to store the data in an external file and then read it in through the data step. The following code will read data in a text file from a web page.

```
ex.data <- read.table(file="https://math.unm.edu/**/**.txt", header=T)
# Use header=T if there is a header
ex.data    # print the data
```

Suppose the folder on your computer containing the data sets is C:\MyFilePath\. To read a csv file, use **read.csv**. Note that to specify a file path in R, you should replace each backslash '\' with either a forward slash '/' or a double backslash '\\'.

```
ex.data <- read.csv(file="C:/MyFilePath/exampledata.csv",header=F)
```

```
# header=F is the default
colnames(ex.data) <- c("y","x1","x2") #add column names
ex.data
```

**Import data with RStudio.** Use the "Import Dataset" dropdown from the "Environment" window (top right panel in RStudio). The import formats are grouped into 3 categories: text data, spreadsheet data, and statistical data.

- import data from webpage

  Select "From Text (readr)", enter URL address, and click "Import".

- import data from local file

  Select "From Text (base)" or "From Excel", select a local file and click "Import"

Luraschi (2021) gives a detailed description of how to import data in RStudio.

**Read R data sets.** This is the easiest method of all—provided someone else has already saved the file as an R data set. All the data sets in SDA are loaded in R package SDAResources. Here is an example to read data *agpop*.

```
library(SDAResources)
data(agpop)   # Load the agpop data set
N <- nrow(agpop)
N
## [1] 3078
head(agpop)
##                    county state acres92 acres87 acres82 farms92 farms87 farms82
## 1 ALEUTIAN ISLANDS AREA     AK  683533  726596  764514      26      27      28
## 2         ANCHORAGE AREA    AK   47146   59297  256709     217     245     223
## 3         FAIRBANKS AREA    AK  141338  154913  204568     168     175     170
## 4            JUNEAU AREA    AK     210     214     127       8       8      12
## 5   KENAI PENINSULA AREA    AK   50810   85712   98035      93     119     137
## 6         AUTAUGA COUNTY    AL  107259  116050  145044     322     388     453
##   largef92 largef87 largef82 smallf92 smallf87 smallf82 region
## 1       14       16       20        6        4        1      W
## 2        9       10       11       41       52       38      W
## 3       25       28       21       12       18       25      W
## 4        0        0        0        5        4        8      W
## 5        9       18       17       12       18       19      W
## 6       25       32       32        8       19       17      S
```

## 1.5 Saving Output

Section 1.4 gave four methods for reading data into R. How do you save the output from the program? Here are several methods that will allow you to save the output or paste it into another document. Section 1.6 will show you how to incorporate R output directly into a LATEX document.

**Use the *sink* function to save console input and output.** First, use *sink* to create a file in local drive. Next, run the R program. Last, close *sink*. Below is an example.

```
sink(file = "C:/MyOutputPath/output.txt",append = TRUE,
     type = c("output","message"))
```

```
cat("boot.samples", sep="\n")   #add name of the R output
#R program
set.seed (244)
B <- 6
n <- 5
boot.samples <- matrix(sample(agpop$acres92, size=B*n, replace=TRUE),B,n)
boot.samples
sink(file = NULL) #close sink
```

The file C:\MyOutputPath\output.txt in the local drive appears as follows:

```
boot.samples
        [,1]      [,2]    [,3]    [,4]    [,5]
[1,]      831  1449976   98142  161724  138986
[2,]    86856   215222  245681  576468   80905
[3,]    96474    45214   51836  284888  302352
[4,]   137426    17392  141260  117768  177189
[5,]   342653   338529   73654  118291  297984
[6,]    50319     2531  223429  710546  216638
```

In some cases you may also want to save the R commands, so you can look at them later. You may use the function *source* to read in an R program with option echo=TRUE to include the R code, so that R output together with R code will be saved in output.txt.

```
sink(file ="MyOutputPath/output.txt",append = TRUE,
     type = c("output","message"))
source("MyInputFilePath/homework3.R", echo=TRUE)
sink(file = NULL)
```

**Convert a matrix to LATEX code in table format.** In case you want to include a matrix as part of a LATEX document, there is an easy way to convert the matrix to LATEX code in table format. The following code, using the **stargazer** package, converts the matrix *boot.samples* to a LATEX table.

```
# save output as latex
library(stargazer)
set.seed (244)
B <- 6
n <- 5
boot.samples <- matrix(sample(agpop$acres92, size=B*n, replace=TRUE),B,n)
# Now use stargazer to produce Latex commands for table
# Add header=FALSE to omit the initial comments in the Latex code output.
stargazer(boot.samples, digits = 2,header=FALSE)
##
## \begin{table}[!htbp] \centering
##   \caption{}
##   \label{}
## \begin{tabular}{@{\extracolsep{5pt}} ccccc}
## \\[-1.8ex]\hline
## \hline \\[-1.8ex]
## $831$ & $1,449,976$ & $98,142$ & $161,724$ & $138,986$ \\
## $86,856$ & $215,222$ & $245,681$ & $576,468$ & $80,905$ \\
## $96,474$ & $45,214$ & $51,836$ & $284,888$ & $302,352$ \\
## $137,426$ & $17,392$ & $141,260$ & $117,768$ & $177,189$ \\
## $342,653$ & $338,529$ & $73,654$ & $118,291$ & $297,984$ \\
## $50,319$ & $2,531$ & $223,429$ & $710,546$ & $216,638$ \\
```

```
## \hline \\[-1.8ex]
## \end{tabular}
## \end{table}
```

**Saving graphs.** R and RStudio both allow you to save a graph using menu options. In standalone R, you can save a graph by choosing "Save as" from the File menu. In RStudio, select the Export dropdown from the plot panel (lower right-panel):

Plots→ Export → Save as Image or Save as PDF.

Alternatively, you can specify files to save the image using functions such as *jpeg*, *png*, and *pdf*, which save the plot in an external file with the designated format. All of these functions have multiple options for sizing and formatting the graph. To save a graph as a pdf file, for example, first open the pdf file, then create the plot, and then close the file:

```
# Open a pdf file
pdf("~/MyOutputPath/rplot.pdf")
# Create a plot
boxplot(acres92/10^6 ~ region, xlab = "Region", ylab = "Millions of Acres",
        data = agpop)
# Close the pdf file
dev.off()
```

Or, to save as a jpeg file,

```
# save as jpeg file
jpeg("~/MyOutputPath/rplot.jpg",width=350,height=350)
boxplot(acres92/10^6 ~ region, xlab = "Region", ylab = "Millions of Acres",
        data = agpop)
dev.off()
```

**Saving a data set.** You can use function *write.table* or *write.csv* to save a data set to an external file. In the following, we will use data *classes* to create a new dataset in a long format called *classeslong*, and use *write.csv* to save the data in file **classeslong.csv**.

```
# read in data classes
data(classes)
# change to a long format by creating a record for each student
# create new data frame with each row repeated as many times as number of students
classeslong<-classes[rep(1:nrow(classes),times=classes$class_size),]
# add column of student ids, goes from 1 to number of students in each class
classeslong$studentid <- sequence(classes$class_size)
nrow(classeslong)
## [1] 647
head(classeslong)
##      class class_size studentid
## 1        1         44         1
## 1.1      1         44         2
## 1.2      1         44         3
## 1.3      1         44         4
## 1.4      1         44         5
## 1.5      1         44         6
# you can save classeslong in your local drive (or other file path) using write.csv
write.csv(classeslong, file="~/classeslong.csv", row.names = FALSE)
```

## 1.6 Integrating R Output into LaTeX Documents

The `Sweave` system within R allows you to embed R code and output within LaTeX documents to generate a `pdf` file that includes narrative, graphics, code (if needed), and the results of the computations in R. The R package `knitr` serves as an engine for `Sweave` and allows you to generate dynamic reports with reproducible research (that is, someone else can produce the same results when given access to the data and code) using R. Help for `knitr` is available on the web (`https://yihui.org/knitr/`) and in the book by Xie (2015). We used the `knitr` package to produce this book.

Let's start with a simple example, using RStudio.

**Step 1.** Open a new `Rnw` script by clicking the "R Sweave" icon under the "New File" icon in the upper left of the main RStudio toolbar. Your first Rnw file will appear as in Figure 1.2. Save the file with extension `Rnw`—for example, you might name this file `test.Rnw`.

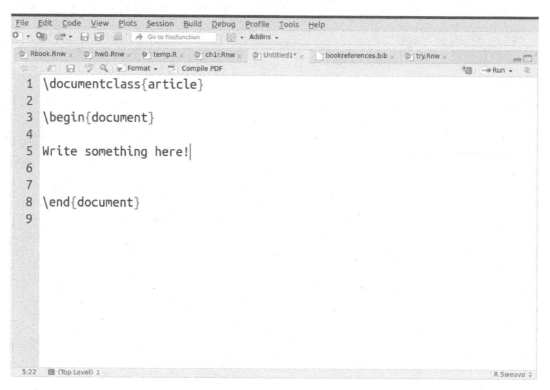

**FIGURE 1.2:** Example of a `Rnw` file.

**Step 2.** Weave the `Rnw` file using `knitr`. You will be asked to install package `knitr` if this is the first time you have used it.
Click Tools → Global options → Sweave → Weave Rnw files using → knitr.

**Step 3.** Click "Compile PDF" to generate a pdf file. The file `test.pdf` will appear in the same directory as `test.Rnw`, along with the LaTeX file, log file, and other auxiliary files.

**Embedding R code within LaTeX.** Sweave allows you to embed R code within LaTeX documents so you can incorporate statistical results in your pdf document. Type your LaTeX document in the program editor window in RStudio. To embed a "chunk" of R code in knitr, start with << ·· >>= (all on one line) and close with @. Below is an illustration (we had to break the first line so it fit on the page, but you should leave it unbroken):

```
<<exercise0, echo=TRUE, size='footnotesize', include=TRUE, fig.width=6,
    fig.height=5, out.width='0.80\\textwidth'>>=
# Insert R code here!
x <- 1:10
y <- seq(from=20,to=2,by=-2)
x
y
x*y
cor(x,y)
# End the chunk with an @
@
```

This example, which has label *exercise0*, includes the following options to customize the R output.

**echo** = TRUE (print source code in the document) or FALSE (do not print source code in the document).

**size** = font size of the code text, if echoed. Options include 'normalsize,' 'footnotesize,' 'large,' and other font sizes in LATEX.

**include** = TRUE if it is desired to include the chunk output in the document. If FALSE, the output is not produced, but the R code is still executed and the plots are generated.

**fig.width** = figure width in inches

**fig.height** = figure height in inches

**out.width** = scaling figures to fit on page (can be inches or cm, or relative to page size). Here, the width is scaled to be 0.8 times the value of the *textwidth* in the LATEX document.

To learn more about chunk options, please refer to `https://yihui.org/knitr/options/#chunk_options`.

Here is what is displayed when we include the code for *exercise0* in the document:

```
# Insert R code here!
x <- 1:10
y <- seq(from=20,to=2,by=-2)
x
##  [1]  1  2  3  4  5  6  7  8  9 10
y
##  [1] 20 18 16 14 12 10  8  6  4  2
x*y
##  [1] 20 36 48 56 60 60 56 48 36 20
cor(x,y)
## [1] -1
# End the chunk with an @
```

Now, try a slightly more complicated example in `hw0.Rnw` (included on the book webpage) to see how to embed R chunks (including graphs) within latex, and compile it. You can see the pdf document that is produced by the example in file `hw0.pdf`.

Many excellent books are available for learning and using LATEX. The books by Talbot (2012), Oetiker et al. (2021), and Wikibooks contributors (2021) may be downloaded free of charge.

## 1.7   Missing Data

Many survey data sets have observations that are missing. For example, the data file
`agpop.csv`, from which the sample used in Chapter 2 is drawn, has missing data. As is
common in data sets intended to be readable by multiple programs, a designated number is
used to indicate that the data value is missing. In this data set, the value "−99" indicates
that the data value is missing. This value must be recoded to the R symbol that indicates
missing data before performing calculations. Otherwise, if, say, you want to calculate the
mean of a variable, a function will treat all the "−99"s as if they were observations with
the value −99 instead of missing values—this could lead to embarrassing results such as
computing a negative value for the average number of acres per farm.

Missing values in R are coded by the symbol NA. If your data set has missing values that
are coded as a number (such as −99), you should recode those to NA before starting your
analysis. Missing data in the R data sets in package **SDAResources** (Lu and Lohr, 2021)
have been recoded to NA, but the data sets for SDA that are in `.csv` format use other
codes, which are described in Appendix A. If you want to read the data set `agpop.csv`
into R from the `.csv` file (instead of loading it from **SDAResources**), you should recode the
missing data as follows:

```
agpop1 <- read.csv(file="~/agpop.csv",header=TRUE)
# some variables have missing values, coded as -99
sum(agpop1==-99)  # 59 missing values altogether
## [1] 59
# look at a row with missing data
agpop1[200,]
##                    county state acres92 acres87 acres82 farms92 farms87 farms82
## 200 SAN FRANCISCO COUNTY    CA       7     -99      19       6       5       7
##      largef92 largef87 largef82 smallf92 smallf87 smallf82 region
## 200         0        0        0        6        5        7      W
agpop2 <- agpop1
# recode missing values to NA
agpop2[agpop1==-99] <- NA
# look at row with missing data again
agpop2[200,]
##                    county state acres92 acres87 acres82 farms92 farms87 farms82
## 200 SAN FRANCISCO COUNTY    CA       7      NA      19       6       5       7
##      largef92 largef87 largef82 smallf92 smallf87 smallf82 region
## 200         0        0        0        6        5        7      W
# count missing values in recoded data
sum(is.na(agpop2))
## [1] 59
```

Different functions in R treat missing data in different ways. For many functions, the treatment of missing values can be specified using the *na.rm* argument. The *sum* function,
for example, will compute the sum of the non-missing values in a vector if you include
`na.rm=TRUE`; otherwise, it returns NA if there are missing values.

```
sum(agpop2$acres87)
## [1] NA
sum(agpop2$acres87,na.rm=TRUE)
## [1] 963466689
```

## 1.8 Summary, Tips, and Warnings

Tables 1.1 and 1.2 describe commonly used R functions for working with data. Table 1.3 lists other functions used in this chapter to read and write data. The `base`, `stats`, and `utils` packages are automatically included when you install R on your system.

**TABLE 1.3**
Functions used for Chapter 1.

| Function | Package | Usage |
| --- | --- | --- |
| library | base | Load an R package that you have installed on your system. You need to load a package each time you start a new R session. |
| sink | base | Send R output to an external file |
| source | base | Read and execute R commands from an external file |
| install.packages | utils | Install an R package |
| data | utils | Load a specified data set |
| read.table | utils | Read data from an external file into a data frame. You can specify the character that separates fields, row and column names, and whether the first line of the file gives the variable names. |
| read.csv | utils | This function is like read.table for an external file that is in comma-delimited (`.csv`) format. |
| write.table | utils | Write an R data set to an external file |
| write.csv | utils | Write an R data set to an external file in `.csv` format |

### Tips and Warnings

- R commands can be tricky, and sometimes the result of using functions on vectors and matrices may not be what you were expecting. When creating new variables or matrices, print out the first few rows or compute summary statistics to check that you have created the object you want.

- Use vector and matrix commands whenever possible when performing calculations with R. Although we occasionally use *for* loops in this book where this will make the code easier to understand, in general, it is much more efficient to perform operations on vectors.

- Many surveys contain missing data. Check how these are coded in the data set, and recode missing values to NA before starting your analysis.

- Sometimes multiple packages contain functions or data sets with the same names. For example, the `sampling` and `survival` packages (the `survival` package is loaded when you load the `survey` package) both have functions named *strata* and *cluster*. If you load `sampling` and then load `survival`, you will get the function from the `survival` package when you call *strata*—the function *strata* from the `sampling` package is masked. To avoid this problem, load the `sampling` package after loading the `survival` package or access the function from `sampling` by typing `sampling::strata`. For this book, we suggest loading the packages in the order given in Section 1.2.

# 2

## Simple Random Sampling

Data from a simple random sample (SRS) can be analyzed using R functions that are designed for data that can be considered as independent and identically distributed, and an SRS can be selected using the R *sample* function. For other types of probability samples, however, you either need to write your own function to account for the survey design or employ functions that have been written by other R users in contributed packages. This chapter reviews how to select a sample and compute estimates from an SRS using functions in base R. It also introduces you to the *srswor* and *srswr* functions from the **sampling** package (Tillé and Matei, 2021) to select an SRS, with or without replacement; and the *svydesign*, *svymean*, and *svytotal* functions from the **survey** package (Lumley, 2020) to calculate the statistics.

All code in this chapter can be found in file `ch02.R` on the book website (see the Preface for the website address). The data sets are available from the book website and in the R package `SDAResources` (Lu and Lohr, 2021). The variables in the data sets are described in Appendix A.

In this and future chapters, load the following three packages before starting the computations. You installed these packages in Chapter 1. In the R console, type:

```
library(survey)
library(sampling)
library(SDAResources)
```

Before calculating statistics, let's first look at how to use R functions to select an SRS from a population.

## 2.1 Selecting a Simple Random Sample

**Example 2.5 of SDA.** *Selecting an SRS from a population.* SDA used a random number table to select an SRS of size 4 from a population of size 10. There are several options for selecting an SRS in R.

**Using the *sample* function in base R.** Base R contains the function *sample* that can be used to select an SRS. We can select an SRS (without replacement) of size 4 from a population of size 10 as follows:

```
# Set the seed for random number generation
set.seed(108742)
# Select an SRS of size n=4 from a population of size N=10 without replacement
srs4 <- sample(1:10, 4, replace = FALSE)
# Print the sample to see
# Can print an object by typing its name, or typing "print(object)"
```

```
srs4
## [1] 1 8 9 5
```

The first line of the code uses the function

```
set.seed(seed)
```

where *seed* is an integer that you supply to the function. If you want to be able to reproduce your sample later, call *set.seed* immediately before calling the function that generates a sample, and record the value of *seed* that you used. You will then get the same sample the next time you call the *sample* function with the same value of *seed*.[1] If you do not call *set.seed* during your R session, the starting point will be generated by the program.

For this example, the *sample* function provided a sample containing units 1, 8, 9, and 5. Running the sample function again, without using *set.seed*, gives a different sample. But when we reset the seed to the original value of 108742, we obtain the first sample $\{1, 8, 9, 5\}$ again.

```
# Run again, without setting a new seed.
sample(1:10, 4, replace = FALSE)
## [1]  9  7  3 10
# Now go back to original seed.
set.seed(108742)
sample(1:10, 4, replace = FALSE)
## [1] 1 8 9 5
```

The *sample* function is called with

```
sample(x,size,replace=FALSE)
```

to select an SRS without replacement of *size* observations from the population in $x$. We sample 4 observations from the population in the vector $[1, 2, \ldots, 10]$.

The *sample* function will also select a simple random sample with replacement by calling it with `replace = TRUE`. Unit "9" appears twice in the following with-replacement sample.

```
# Using the sample function to select an SRS with replacement
set.seed(101)
srswr4 <- sample(1:10, 4, replace = TRUE)
srswr4
## [1] 9 9 7 1
```

**Using the *srswor* or *srswr* function from the sampling package.** An alternative is to use function *srswor* or *srswr* to select an SRS without or with replacement, respectively. These are in the `sampling` package (Tillé and Matei, 2021), which you installed in Chapter 1. Now load the package and select a sample by calling:

```
srswor(n,N)
```

where $n$ is the sample size and $N$ is the population size.

```
# Load the sampling package if you have not already done so.
# library(sampling)
set.seed(1329)
# Select an SRS of size n=4 from a population of size N=10 without replacement.
s1<-srswor(4,10)
# List the units in the sample (the population units having s1=1).
```

---

[1]Samples generated by the *sample* function in R versions 3.6.0 and later, however, will differ from samples generated by earlier versions of R because the *sample* function was revised in version 3.6.0 to fix a bug.

```
s1
## [1] 0 0 1 1 1 0 0 0 1 0
(1:10)[s1==1]
## [1] 3 4 5 9
```

The function *srswor* returns of vector of length $N$, with ones in the positions of the units selected for the sample. The sample in *s1* consists of units 3, 4, 5, and 9.

Function *srswr* for drawing a with-replacement sample is similar, but returns a vector containing the number of times each unit is in the sample.

```
# Select an SRS of size n=4 from a population of size N=10 with replacement.
set.seed(35882)
s2<-srswr(4,10)
# the selected units are 2 and 9
s2
## [1] 0 2 0 0 0 0 0 0 2 0
(1:10)[s2!=0]
## [1] 2 9
# number of replicates, units 2 and 9 both appear twice
s2[s2!=0]
## [1] 2 2
# can use the getdata function to extract the sample from data frame with population
popdf<-data.frame(popid=1:10)
getdata(popdf,s2)
## [1] 2 2 9 9
```

The *getdata* function in the **sampling** package extracts the sample from the population listing. For this example, since units 2 and 9 are each selected twice, *getdata* repeats each of these units twice.

Note that the *sample* function has the sample size as the second argument, while the *srswor* and *srswr* functions have the sample size as the first argument. When you type `srswr(4,10)`, R assigns the values to variables in the order given in the function definition (see Usage in the help file for the function). If you want to assign values in a different order (or even if you want to use the same order but want to be able to read the call easily), name the variables when calling the function. All of the following are equivalent:

```
# Call a function using variable names
set.seed(35882)
srswr(4,10)
## [1] 0 2 0 0 0 0 0 0 2 0
set.seed(35882)
srswr(n=4,N=10)
## [1] 0 2 0 0 0 0 0 0 2 0
set.seed(35882)
srswr(N=10,n=4)
## [1] 0 2 0 0 0 0 0 0 2 0
```

**Example 2.6 of SDA.** In SDA, the sample in `agsrs.csv` was selected from the population `agpop.csv` using random numbers generated in a spreadsheet. In the following, we show how to use R code to select a different SRS of size 300 from the 3078 counties in data set *agpop*. The function *getdata* is used to extract the sampled units from the population data, and the first 6 observations are printed.

```
# Select a different sample of size 300 from agpop
# Load the SDAResources package containing all the data in SDA book
```

```
# library(SDAResources)  # we comment this since we already loaded the package
data(agpop)  # Load the data set agpop
N <- nrow(agpop)
N  # 3078 observations
## [1] 3078
# Select an SRS of size n=300 from agpop
set.seed(8126834)
index <- srswor(300,N)
# each unit k is associated with index 1 or 0, with 1 indicating selection
index[1:10]
## [1] 0 0 0 1 0 0 0 0 0 0
# agsrs2 is an SRS with size 300 selected from agpop
# extract the sampled units from the data frame containing the population
agsrs2 <- getdata(agpop,index)
agsrs2 <- agpop[(1:N)[index==1],] # alternative way to extract the sampled units
head(agsrs2)
##            county state acres92 acres87 acres82 farms92 farms87 farms82
## 4     JUNEAU AREA    AK     210     214     127       8       8      12
## 30  DE KALB COUNTY    AL  210733  213440  221502    1894    2047    2228
## 38    HALE COUNTY    AL  167583  154581  179618     382     441     481
## 46     LEE COUNTY    AL   67962   79836  100949     336     402     407
## 50 MADISON COUNTY    AL  224370  235478  292873     871     977    1101
## 62 RUSSELL COUNTY    AL  112620  143568  141048     213     276     314
##     largef92 largef87 largef82 smallf92 smallf87 smallf82 region
## 4          0        0        0        5        4        8      W
## 30        13        5        6      114      133      168      S
## 38        38       33       39       12       22       17      S
## 46        10       10       20       15       22       20      S
## 50        59       59       61       46       76       89      S
## 62        25       30       33       14       14       25      S
```

The functions *sample*, *srswor*, and *srswr* select a sample but do not provide sampling weights. After drawing the sample, you need to create a variable of sampling weights for the sampled units. For *agsrs2*, the weight variable *sampwt* has the value 3078/300 for all units.

```
# Create the variable of sampling weights
n <- nrow(agsrs2)
agsrs2$sampwt <- rep(3078/n,n)
# Check that the weights sum to N
sum(agsrs2$sampwt)
## [1] 3078
```

## 2.2  Computing Statistics from a Simple Random Sample

In this section, we look at two ways of computing estimates from an SRS: by using standard functions in R to calculate estimates from the formulas and by using functions in the **survey** package. We shall use the **survey** package to calculate estimates for the other chapters of this book, but show how to compute estimates for an SRS using the formulas so you can see that the two methods give the same numbers. Since R is a highly flexible language, you can also write your own functions to compute estimates using the formulas.

**Examples 2.6, 2.7, and 2.11 of SDA.** This section analyzes variables in data set `agsrs.csv`, described in Example 2.6 of SDA and in Appendix A of this book. The primary response variable for these examples is *acres92* (acreage devoted to farms in 1992).

First, we draw a histogram of the variable *acres92*, shown in Figure 2.1. The optional argument `breaks=20` tells R to use 20 bins for the histogram. The *col* (specifies color of bars), *xlab* (gives label for the x-axis), and *main* (specifies the title for the graph) arguments are also optional but make the picture look nicer.

```
# Draw a histogram
hist(agsrs$acres92,breaks=20,col="gray",xlab="Acres devoted to farms, 1992",
    main="Histogram: Number of acres devoted to farms, 1992")
```

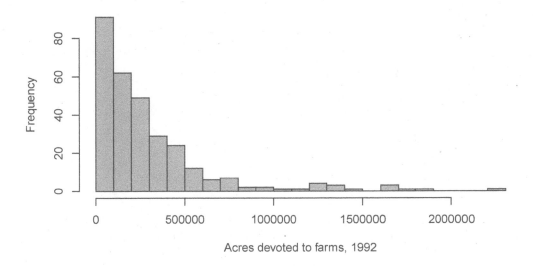

**Histogram: Number of acres devoted to farms, 1992**

FIGURE 2.1: Histogram of farm acreage in 1992 (data *agsrs*).

**Using the formulas in SDA.** Statistics from an SRS can be computed using functions supplied in the base R package with the formulas in SDA. Functions such as *t.test* will compute confidence intervals without a finite population correction (fpc).

```
# Base R functions such as t.test will calculate statistics for an SRS,
# but without the fpc.
t.test(agsrs$acres92)
##
##  One Sample t-test
##
## data:  agsrs$acres92
## t = 14.975, df = 299, p-value < 2.2e-16
## alternative hypothesis: true mean is not equal to 0
## 95 percent confidence interval:
##  258749.6 337044.5
## sample estimates:
## mean of x
##    297897
```

The confidence interval from the *t.test* function is wider than that in Example 2.11 of SDA, however. Standard errors and confidence intervals that incorporate the fpc can be calculated directly with the formulas. If you are familiar with writing functions in R, you could also write your own function to do these calculations.

```
# Calculate the statistics by direct formula
n <- length(agsrs$acres92)
ybar <- mean(agsrs$acres92)
ybar
## [1] 297897
hatvybar<-(1-n/3078)*var(agsrs$acres92)/n
seybar<-sqrt(hatvybar)
seybar
## [1] 18898.43
# Calculate confidence interval by direct formula using t distribution
Mean_CI <- c(ybar - qt(.975, n-1)*seybar, ybar + qt(.975, n-1)*seybar)
names(Mean_CI) <- c("lower", "upper")
Mean_CI
##    lower    upper
## 260706.3 335087.8
# To obtain estimates for the population total,
# multiply each of ybar, seybar, and Mean_CI by N = 3078
seybar*3078
## [1] 58169381
Mean_CI*3078
##       lower       upper
##   802453859 1031400361
# Calculate coefficient of variation of mean
seybar/ybar
## [1] 0.06343948
```

**Using functions in the *survey* package.** Most of the statistics discussed in Chapter 2 can be computed in R using functions *svydesign*, *svymean*, and *svytotal* from the **survey** package (Lumley, 2020).

The data set *agsrs* does not contain a variable of sampling weights, so we need to create one. Also define the variable *lt200k*, which takes on value 1 if *acres92* < 200,000 and the value 0 if *acres92* ≥ 200,000. The mean of variable *lt200k* estimates the proportion of farms that have fewer than 200,000 acres.

```
# Create the variable of sampling weights
n <- nrow(agsrs)
agsrs$sampwt <- rep(3078/n,n)
# Create variable lt200k
agsrs$lt200k <- rep(0,n)
agsrs$lt200k[agsrs$acres92 < 200000] <- 1
# look at the first 10 observations with column 3 (acres92) and column 17 (lt200k)
agsrs[1:10,c(3,17)]
##    acres92 lt200k
## 1   175209      1
## 2   138135      1
## 3    56102      1
## 4   199117      1
## 5    89228      1
## 6    96194      1
## 7    57253      1
```

```
## 8    210692      0
## 9     78498      1
## 10   219444      0
```

Now specify the survey design with function *svydesign*. The function has numerous optional arguments, so we call it using the variable names in the arguments. The arguments used for an SRS are:

**id** In general survey designs, *id* specifies the cluster identifiers. For an SRS we use `id = ~1`, which tells *svydesign* that there is no clustering. The tilde (~) is used by R to specify a formula, and the syntax for formulas will be discussed in later chapters.

**weights** Names the variable in the data frame that contains the sampling weights. The weights argument can actually be omitted for calculating means in an SRS (the function will calculate weights from the *fpc* argument if it is supplied, or set all weights equal to 1 if neither *weights* nor *fpc* is included), but it is good to get in the habit of using a weight variable, so we include it here.

**fpc** Information for calculating the finite population correction. For an SRS, we can use `fpc = rep(N, n)` with $N$ the population size and $n$ the sample size. For *agsrs*, $N = 3078, n = 300$, and `fpc = rep(3078, 300)`.

**data** Name of the data frame containing the variables to be analyzed.

```
# If you did not load the survey package at the beginning of the chapter, do it now
# library(survey)
# Specify the survey design.
# This is an SRS, so the only design features needed are the weights
#    or information used to calculate the fpc.
dsrs <- svydesign(id = ~1, weights = ~sampwt, fpc = rep(3078,300), data = agsrs)
dsrs
## Independent Sampling design
## svydesign(id = ~1, weights = ~sampwt, fpc = rep(3078, 300), data = agsrs)
```

When we print *dsrs*, we are told that this is an "Independent Sampling design"—that is, there is no stratification or clustering. We will come back to the function *svydesign* in later chapters as we encounter other survey designs.

Now that the survey design has been specified, we can calculate estimated means and totals using the *svymean* and *svytotal* functions. For each of these, the first argument contains the name(s) of the variable(s) to be analyzed, and the second argument is the name of the design object that was created by *svydesign*. The function *confint* will construct a 95% confidence interval (you can specify other confidence levels with the optional *level* argument, but we omit this since we always use 95% intervals in this book) using a $t$ distribution with $df$ degrees of freedom. If you do not specify the df, a normal distribution will be used for the confidence intervals. For an SRS, the $t$ distribution has $n - 1$ df, where $n$ is the sample size.

```
# Calculate the mean for acres92 and its standard error using the svymean function.
smean <- svymean(~acres92,dsrs)
smean
##              mean    SE
## acres92 297897 18898
# Use the confint function to compute a 95% confidence interval from
# the information in smean, df = n-1 = 300-1 = 299
confint(smean, df=299)
##             2.5 %   97.5 %
```

```
## acres92 260706.3 335087.8
# Repeat these steps with the svytotal function to obtain estimated totals.
stotal <- svytotal(~acres92,dsrs)
stotal
##               total        SE
## acres92 916927110 58169381
confint(stotal, df=299)
##               2.5 %       97.5 %
## acres92 802453859 1031400361
# Calculate the CV of the mean
SE(smean)/coef(smean)
##             acres92
## acres92 0.06343948
# or
smean<-as.data.frame(smean)
smean[[2]]/smean[[1]]
## [1] 0.06343948
```

You can analyze multiple variables at a time by putting them in a formula. The following code estimates the population means for variables *acres92* and *lt200k*.

```
# Estimate population means for multiple variables
agsrs_means <- svymean(~acres92+lt200k,dsrs)
agsrs_means
##                 mean            SE
## acres92 297897.05 18898.4344
## lt200k        0.51        0.0275
confint(agsrs_means, df=299)
##                   2.5 %           97.5 %
## acres92 2.607063e+05 3.350878e+05
## lt200k  4.559508e-01 5.640492e-01
```

**Estimating proportions from an SRS.** For a binary numeric variable (taking on values 0 or 1), the estimated proportion is the mean of the variable, and the proportion of the population having $lt200k = 1$ is the mean of variable *lt200k*. From the above output, we can see that the value of $\hat{p} = 0.51$ is the estimated proportion where *lt200k* takes on the value 1. The standard error is 0.0275, and a 95% confidence interval of $p$ is $[0.456, 0.564]$.

Sometimes you want to estimate the proportion of the population that falls in each of multiple categories. The variable *region* in the *agsrs* data describes the census region for each county in the sample and takes on the values "NE," "NC," "S," and "W". Running *svymean* with the variable *region* gives the estimated proportion in each category.

```
# Analyzing a categorical variable that is coded as characters
# First, display the category names and counts
table(agsrs$region)
##
##  NC  NE   S   W
## 107  24 130  39
# Find the estimated proportions in each category
region_prop <- svymean(~region,dsrs)
region_prop
##                 mean      SE
## regionNC 0.35667 0.0263
## regionNE 0.08000 0.0149
## regionS  0.43333 0.0272
```

```
## regionW   0.13000 0.0185
confint(region_prop,df=299)
##                2.5 %     97.5 %
## regionNC 0.30487557 0.4084578
## regionNE 0.05066780 0.1093322
## regionS  0.37975605 0.4869106
## regionW  0.09363889 0.1663611
region_total <- svytotal(~region,dsrs)
region_total
##              total      SE
## regionNC 1097.82 81.005
## regionNE  246.24 45.878
## regionS  1333.80 83.799
## regionW   400.14 56.872
confint(region_total,df=299)
##                2.5 %     97.5 %
## regionNC  938.4070 1257.2330
## regionNE  155.9555  336.5245
## regionS  1168.8891 1498.7109
## regionW   288.2205  512.0595
```

**Numeric and categorical variables.** Numeric variables are variables for which you want to calculate statistics such as means (for example, *acres92* is a numeric variable). Categorical variables are those for which the values represent categories. Region is a categorical variable. We want to estimate the proportion of the population in each region, but we cannot calculate an "average" region. Here, *region* is automatically recognized as a categorical variable because it contains characters other than numbers.

Some surveys code categories as numbers; be careful to treat such variables as categorical rather than numeric. For example, a survey variable *haircolor* might take values 1–6, where 1 represents black, 2 represents brown, 3 represents blond, 4 represents red, 5 represents bald, and 6 represents other. You can calculate the mean for the variable *haircolor*, but it has no meaning. If you want to estimate the population proportion with each hair color, you can either (1) define binary variables for each category, for example, *redhair* = 1 if *haircolor* = 4 and 0 otherwise and find the mean of each variable, or (2) declare the variable *haircolor* to be categorical.

In R, you specify that a variable is categorical with the *factor* function. You can either declare the variable to be a factor variable in the data set or in the function call of *svymean*.

```
# Analyzing a categorical variable that is coded as numbers
# First, analyze lt200k as a numeric variable (works only if all values are 0 or 1)
# This gives the mean of variable lt200k, which is the proportion with lt200k = 1.
svymean(~lt200k,dsrs)
##          mean      SE
## lt200k 0.51 0.0275
# Now, analyze lt200k as a factor variable. This gives the proportion
# in each category
svymean(~factor(lt200k),dsrs)
##                   mean      SE
## factor(lt200k)0 0.49 0.0275
## factor(lt200k)1 0.51 0.0275
```

## 2.3   Additional Code for Exercises

This section contains additional code and references to functions used in three of the exercises in Chapter 2 of SDA. Some of these make use of advanced features of R; you should skip this section if you are new to R.

**Exercise 2.27 of SDA** asks you to estimate the sampling distribution of $\bar{y}$ by repeatedly taking samples of size $n$ with replacement from the sample in *agsrs*, where $y$ is the variable *acres92*. The following code constructs the histogram of the statistics from the bootstrap replicates that is shown in Figure 2.2. We use the *apply* function to calculate the mean value of *acres92* from each bootstrap replicate.

```
# Calculating bootstrap means for Exercise 2.27 in SDA
set.seed (244)
B = 1000
n = length(agsrs$acres92)
boot.samples = matrix(sample(agsrs$acres92, size = B * n, replace = TRUE),B, n)
boot.statistics = apply(boot.samples, 1, mean)
hist(boot.statistics, main="Estimated Sampling Distribution of ybar",
    xlab="Mean of acres92 from Bootstrap Replicate",
    col="gray",border="white")
```

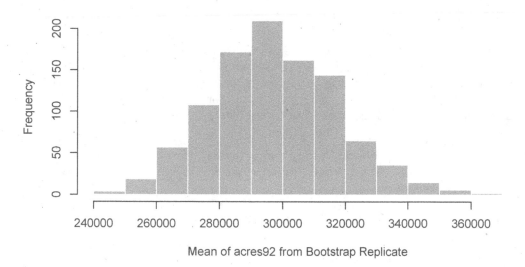

**FIGURE 2.2:** Histogram of means from bootstrap replicates.

**Exercise 2.32 of SDA.** The function *srswor1* in the **sampling** package will select an SRS using the algorithm in this exercise.

**Exercise 2.34 of SDA.** The function *binom.test* will calculate Clopper-Pearson confidence intervals for an SRS. For complex surveys, the function *svyciprop* in the **survey** package will calculate a variety of asymmetric confidence intervals for a proportion (see Section 3.4).

## 2.4   Summary, Tips, and Warnings

Table 2.1 lists the major functions used in this chapter to select or analyze data from an SRS. The base, graphics, stats, and utils packages are automatically included when you install R on your system.

**TABLE 2.1**
Functions used for Chapter 2.

| Function | Package | Usage |
|---|---|---|
| sample | base | Select a simple random sample with or without replacement |
| mean | base | Calculate the mean of a vector |
| var | base | Calculate the variance of a vector |
| table | base | Compute the counts in each category of a vector |
| factor | base | Convert a vector to a factor object |
| set.seed | base | Initialize the seed for the random number generator |
| hist | graphics | Draw a histogram of data from an SRS |
| qt | stats | qt(.975,df) calculates the 0.975 quantile of a $t$ distribution with the specified df |
| t.test | stats | Calculate a $t$ confidence interval for an SRS (assumed to be with-replacement) |
| confint | stats | Calculate confidence intervals |
| data | utils | Loads the data set enclosed in parentheses; `data()` lists the available data sets |
| srswor | sampling | Select a simple random sample without replacement |
| srswr | sampling | Select a simple random sample with replacement |
| getdata | sampling | Extract the sampled data from a data frame containing the population units |
| svydesign | survey | Specify the survey design, for example, SRS |
| svymean | survey | Calculate mean and standard error of mean (if the variable is numeric), or proportion in each category (if variable is categorical) |
| svytotal | survey | Calculate total and standard error of total |

**Tips and Warnings**

- If you want to be able to draw the same sample from a population at a later date, use the *set.seed* function before calling the sample-generating function.

- Although the *weights* argument is optional in the *svydesign* function for an SRS, it is a good practice to include it for computing estimates in *svymean* or *svytotal*. Most data sets from complex surveys come with weights that must be used to compute estimates correctly.

- When computing weights for an SRS, check that the sum of the weights equals the population size.

- If you are analyzing categorical variables with the *svymean* or *svytotal* functions, make sure to declare them to be *factor* variables. Otherwise, the functions will treat them as numeric variables and calculate the mean instead of computing the proportion in each category.

# 3

## Stratified Sampling

In a stratified random sample, the population is divided into subgroups called strata. An SRS is selected independently from each stratum. In this chapter, we look at methods for allocating and selecting stratified random samples using functions in base R and the **sampling** package (Tillé and Matei, 2021). We then discuss the usage of functions *svydesign*, *svymean*, *svytotal*, and *svyby* from the **survey** package (Lumley, 2020) in a stratified random sample.

All code in this chapter can be found in file `ch03.R` on the book website. As always, load the packages **survey**, **sampling**, and **SDAResources** before starting your work.

### 3.1 Allocation Methods

Data *agpop* contains a stratum variable *region* that describes the census region for each county in the population and takes on the values North Central (NC), Northwest (NE), South (S), and West (W). The following code calculates the population counts ($N_h$) for the variable *region* with the *table* function.

```
data(agpop)  # load the data set
names(agpop) # list the variable names
## [1] "county"   "state"    "acres92"  "acres87"  "acres82"  "farms92"
## [7] "farms87"  "farms82"  "largef92" "largef87" "largef82" "smallf92"
## [13] "smallf87" "smallf82" "region"
head(agpop)  # take a look at the first 6 obsns
##                     county state acres92 acres87 acres82 farms92 farms87 farms82
## 1 ALEUTIAN ISLANDS AREA    AK    683533  726596  764514      26      27      28
## 2          ANCHORAGE AREA  AK     47146   59297  256709     217     245     223
## 3          FAIRBANKS AREA  AK    141338  154913  204568     168     175     170
## 4             JUNEAU AREA  AK       210     214     127       8       8      12
## 5   KENAI PENINSULA AREA   AK     50810   85712   98035      93     119     137
## 6         AUTAUGA COUNTY   AL    107259  116050  145044     322     388     453
##   largef92 largef87 largef82 smallf92 smallf87 smallf82 region
## 1       14       16       20        6        4        1      W
## 2        9       10       11       41       52       38      W
## 3       25       28       21       12       18       25      W
## 4        0        0        0        5        4        8      W
## 5        9       18       17       12       18       19      W
## 6       25       32       32        8       19       17      S
nrow(agpop)  #number of rows, 3078
## [1] 3078
unique(agpop$region) # take a look at the four regions, NC, NE, S, W
## [1] "W"  "S"  "NE" "NC"
table(agpop$region)  # number of counties in each stratum
```

DOI: 10.1201/9781003228196-3

```
##
##   NC   NE    S    W
## 1054  220 1382  422
```

We can use the information about *region* to allocate a stratified sample.

**Proportional allocation.** With proportional allocation, the stratum sample sizes are proportional to the population stratum sizes $N_h$. A proportional allocation is easy to calculate in R; simply multiply $N_h/N$ by the desired sample size. For example, region NC has 1054 counties and the population has 3078 counties. For a sample with $n = 300$, proportional allocation will select $300 * 1054/3078 = 103$ counties from region NC. The values in *propalloc* are fractions, so we round these to the nearest integers to obtain the sample size.

```
popsize <- table(agpop$region)
propalloc <- 300*popsize/sum(popsize)
propalloc
##
##        NC        NE         S         W
## 102.7290   21.4425  134.6979   41.1306
# Round to nearest integer
propalloc_int <- round(propalloc)
propalloc_int
##
##  NC  NE   S   W
## 103  21 135  41
sum(propalloc_int) # check that stratum sample sizes sum to 300
## [1] 300
```

**Neyman allocation.** For Neyman allocation, you need to provide additional information about the stratum variances. Sometimes you have information about a variable that is related to key survey responses from the sampling frame, or sometimes you have information on variances from a pilot study or from similar surveys that have been done. In other cases, you may need to make a conjecture about the stratum variances.

In the following example, we assume that the survey planner does not have the true population variances available, and we enter conjectures for the relative variances of the strata. For example, the variance in the West is set at twice the variance for the South. Using the *popsize* vector that was calculated in the previous code, we have:

```
stratvar <- c(1.1,0.8,1.0,2.0)
# Make sure the stratum variances in stratvar are in same
#   order as the table in popsize
neymanalloc <- 300*(popsize*sqrt(stratvar))/sum(popsize*sqrt(stratvar))
neymanalloc
##
##        NC        NE         S         W
## 101.07640  17.99204 126.36327  54.56828
neymanalloc_int <- round(neymanalloc)
neymanalloc_int
##
##  NC  NE   S   W
## 101  18 126  55
sum(neymanalloc_int)
## [1] 300
```

**Optimal allocation.** Optimal allocation can be done similarly by defining costs or relative costs for sampling in each stratum.

```
relcost <- c(1.4,1.0,1.0,1.8)
# Make sure the relative costs in relcost are in same
# order as the table in popsize
optalloc <- 300*(popsize*sqrt(stratvar/relcost))/sum(popsize*sqrt(stratvar/relcost))
optalloc
##
##        NC        NE         S         W
##  94.75776  19.95766 140.16833  45.11626
optalloc_int <- round(optalloc)
optalloc_int
##
##  NC NE   S   W
##  95 20 140  45
sum(optalloc_int)
## [1] 300
```

Table 3.1 summarizes the results of these three allocation methods for the *agpop* population. Of course, the Neyman and optimal allocations are only optimal under the assumed variances and costs used in the calculations. If those variances or costs are wrong, then these allocations will not be optimal for the variable of interest. And an allocation that is optimal for one response variable might not be optimal for another.

**TABLE 3.1**
Proportional, Neyman, and optimal allocation for the four regions.

| Number of Counties in Stratum | NC | NE | S | W | Total |
|---|---|---|---|---|---|
| Population | 1054 | 220 | 1382 | 422 | 3078 |
| Sample with proportional allocation | 103 | 21 | 135 | 41 | 300 |
| Sample with Neyman allocation | 101 | 18 | 126 | 55 | 300 |
| Sample with optimal allocation | 95 | 20 | 140 | 45 | 300 |

**Other allocation methods.** The sample sizes specified by the proportional, Neyman, and optimal methods are just guidelines. You can set the stratum sample sizes to any values that meet your research needs. For example, if you want to have high precision for comparing stratum means, you may want to select the same number of observations from each stratum.

There are other functions in R that you can use for allocation with stratified data such as functions *strAlloc* from the `PracTools` package (Valliant et al., 2020) and *optiallo* from the `optimStrat` package (Bueno, 2020). The package `SamplingStrata` (Barcaroli, 2014; Barcaroli et al., 2020) provides R functions for determining the optimal stratification and allocation that will achieve predetermined precisions for multiple $y$ variables. For example, you can use the package to design a stratification that will ensure that the coefficients of variation for five key variables do not exceed 0.05. The package `stratification` (Baillargeon and Rivest, 2011; Rivest and Baillargeon, 2017) contains functions for determining stratum boundaries when the stratifying variable is continuous.

## 3.2   Selecting a Stratified Random Sample

The sample in Example 3.2 of SDA was selected using a spreadsheet but let's look at how to select a similar sample using R, with the *sample* and *strata* functions (this will, of course, give a different sample than obtained in Example 3.2 of SDA).

**Using the *sample* function in base R.** As we have discussed in Chapter 2, the *sample* function can be used to select an SRS. To select a stratified random sample, we select an SRS independently from each stratum.

Data `agpop.csv` contains a stratum variable *region* that describes the census region for each county in the population. In the following example, we use the proportional allocation from Table 3.1 to divide the $n = 300$ units among the four strata, that is, selecting 103, 21, 135, and 41 counties from regions NC, NE, S, and W, respectively.

```
# Select an SRS without replacement from each region with proportional allocation
# with total size n=300
regionname <- c("NC","NE","S","W")
# Make sure sampsize has same ordering as regionname
sampsize <- c(103,21,135,41)
# Set the seed for random number generation
set.seed(108742)
index <- NULL
for (i in 1:length(sampsize)) {
  index <- c(index,sample((1:N)[agpop$region==regionname[i]],
                    size=sampsize[i],replace=F))
}
strsample<-agpop[index,]
# Check that we have the correct stratum sample sizes
table(strsample$region)
##
##  NC  NE   S   W
## 103  21 135  41
# Print the first six rows of the sample to see
strsample[1:6,]
##                          county state acres92 acres87 acres82 farms92 farms87
## 1316                     ISANTI COUNTY    MN  131563  142998  153003     680     817
## 2034                   DEFIANCE COUNTY    OH  196759  206905  210781     830     987
## 864                    ATCHISON COUNTY    KS  245099  233619  234730     686     694
## 553                  DES MOINES COUNTY    IA  192467  210843  224770     681     753
## 1738                       DUNN COUNTY    ND 1352738 1358843 1397141     650     733
## 1325 LAKE OF THE WOODS COUNTY    MN  103665  118959  119296     176     222
##      farms82 largef92 largef87 largef82 smallf92 smallf87 smallf82 region
## 1316     947       18       14        8       14       26       34     NC
## 2034    1033       25       20       18       40       50       50     NC
## 864      768       55       42       41       48       48       65     NC
## 553      815       33       30       24       56       56       72     NC
## 1738     697      358      368      361       19       13       34     NC
## 1325     230       30       35       26        4        4        1     NC
```

This simple code used a *for* loop to select an SRS from each stratum (defined by the subset having *region* equal to the stratum name) in turn; alternatively, one could use the *tapply* function, or write a custom R function, to do this without looping. The vector containing the sample sizes must be in the same order as the vector giving the stratum names.

**Using the *strata* function from the sampling package.** An alternative is to use function *strata* to select a stratified random sample. This function is in the `sampling` package (Tillé and Matei, 2021), which you installed in Chapter 1.

First, sort the data by the stratification variable *region* before selecting the sample. Next, call the *strata* function with sorted data *agpop2* and the stratification variable *region* with first argument `agpop2` and second argument `stratanames="region"`. You can also use a vector of variables to define the strata, such as `stratanames=c("A","B")`, if the strata are formed from multiple variables. Add the information on number of counties to be selected within each stratum by `size=c(103,21,135,41)` in the *strata* function. Finally, choose the method to select the sample within each stratum; for this chapter we use either SRS without replacement (`method="srswor"`) or SRS with replacement (`method="srswr"`).

```
# Sort the population by stratum
agpop2<-agpop[order(agpop$region),]
# Use the strata function to select the units for the sample
# Make sure size argument has same ordering as the stratification variable
index2<-strata(agpop2,stratanames=c("region"),size=c(103,21,135,41),
            method="srswor")
table(index2$region)  # look at number of counties selected within each region
##
##  NC  NE   S   W
## 103  21 135  41
head(index2)
##     region ID_unit      Prob Stratum
## 2       NC       2 0.09772296       1
## 9       NC       9 0.09772296       1
## 27      NC      27 0.09772296       1
## 36      NC      36 0.09772296       1
## 42      NC      42 0.09772296       1
## 43      NC      43 0.09772296       1
strsample2<-getdata(agpop2,index2) # extract the sample
head(strsample2)
##               county state acres92 acres87 acres82 farms92 farms87 farms82
## 526     ADAMS COUNTY    IA  239800  243607  254071     643     688     737
## 533    BREMER COUNTY    IA  236668  235086  250402    1058    1140    1287
## 551   DECATUR COUNTY    IA  261494  278714  300684     648     715     769
## 560   FREMONT COUNTY    IA  302352  308796  306786     596     719     771
## 566    HARDIN COUNTY    IA  332358  337990  355823     986    1065    1208
## 567 HARRISON COUNTY    IA  399155  387190  408601     919    1024    1192
##     largef92 largef87 largef82 smallf92 smallf87 smallf82 region ID_unit
## 526       38       32       21       40       50       33     NC       2
## 533       25       18       11       96      116      109     NC       9
## 551       52       54       56       20       34       37     NC      27
## 560       91       72       51       37       59       50     NC      36
## 566       56       36       42       90      115      132     NC      42
## 567       88       62       51       60       60       66     NC      43
##          Prob Stratum
## 526 0.09772296       1
## 533 0.09772296       1
## 551 0.09772296       1
## 560 0.09772296       1
## 566 0.09772296       1
## 567 0.09772296       1
```

The data frame *index2* contains the stratum variables, the identifiers of the units selected to be in the sample, and the inclusion probability for each unit in the sample. The function *getdata* then extracts the sampled units from the population data.

The *strata* function gives the inclusion probabilities for the sample units but not the weights. You can calculate the sampling weights by taking the reciprocal of the inclusion probabilities. When calculating weights for a stratified random sample, always check that the weights sum to the stratum population sizes. If they do not sum to the stratum population sizes, you have made a mistake somewhere in the weight calculations.

```
# Calculate the sampling weights
# First check that no probabilities are 0
sum(strsample2$Prob<=0)
## [1] 0
strsample2$sampwt<-1/strsample2$Prob
# Check that the sampling weights sum to the population sizes for each stratum
tapply(strsample2$sampwt,strsample2$region,sum)
##   NC   NE    S    W
## 1054  220 1382  422
```

## 3.3   Computing Statistics from a Stratified Random Sample

**Examples 3.2 and 3.6 of SDA.**   As in Chapter 2, function *svydesign* from the `survey` package can be used to enter the stratified random sample information, and functions *svymean* and *svytotal* will calculate estimated means and totals from a stratified random sample. The data set *agstrat* is a stratified random sample taken from the population data *agpop* with proportional allocation. First, let's look at the data.

```
data(agstrat)
names(agstrat)  # list the variable names
##  [1] "county"   "state"    "acres92"  "acres87"  "acres82"  "farms92"
##  [7] "farms87"  "farms82"  "largef92" "largef87" "largef82" "smallf92"
## [13] "smallf87" "smallf82" "region"   "rn"       "strwt"
agstrat[1:6,1:8] # take a look at the first 6 obsns from columns 1 to 8
##      county state acres92 acres87 acres82 farms92 farms87 farms82
## 1 PIERCE C    NE  297326  332862  319619     725     857     865
## 2 JENNINGS    IN  124694  131481  139111     658     671     751
## 3 WAYNE CO    OH  246938  263457  268434    1582    1734    1866
## 4 VAN BURE    MI  206781  190251  197055    1164    1278    1464
## 5  OZAUKEE    WI   78772   85201   89331     448     483     527
## 6 CLEARWAT    MN  210897  229537  213105     583     699     693
nrow(agstrat)   # number of rows, 300
## [1] 300
unique(agstrat$region) # take a look at the four regions, NC, NE, S, W
## [1] "NC" "NE" "S"  "W"
table(agstrat$region)   # number of counties in each stratum
##
##  NC  NE   S   W
## 103  21 135  41
# check that the sum of the weights equals the population size
sum(agstrat$strwt)  #3078
## [1] 3078
```

Figure 3.1 gives a boxplot for variable *acres92* (scaled to millions of acres). We can see that the West region has the highest median and largest variability, while the Northeast region has the lowest median and smallest variability. Note that we can use the *boxplot* function in the following code because an SRS is taken within each stratum (and, because of proportional allocation, the sample is approximately self-weighting); for other designs, one should incorporate the weights into the plot as shown in Chapter 7.

```
boxplot(acres92/10^6 ~ region, xlab = "Region", ylab = "Millions of Acres",
        data = agstrat)
# notice the large variability in western region
```

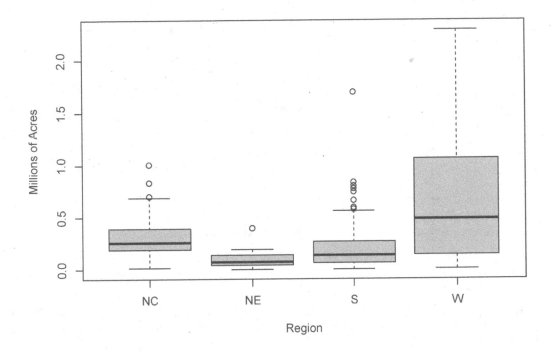

**FIGURE 3.1:** Boxplot of 1992 acreage by region (data *agstrat*).

Now let's calculate the estimates from the stratified random sample. We use function *svydesign* to input the design information and functions *svymean* and *svytotal* to calculate the survey statistics. The following gives the code to find estimates for data *agstrat* along with the output of the statistics calculated. These estimates were given in Examples 3.2 and 3.6 of SDA.

First, we set up the information for the survey design.

```
# create a variable containing population stratum sizes, for use in fpc (optional)
# popsize_recode gives popsize for each stratum
popsize_recode <- c('NC' = 1054, 'NE' = 220, 'S' = 1382, 'W' = 422)
# next statement substitutes 1054 for each 'NC', 220 for 'NE', etc.
agstrat$popsize <- popsize_recode[agstrat$region]
table(agstrat$popsize)  #check the new variable
##
```

```
##  220  422 1054 1382
##   21   41  103  135
# input design information for agstrat
dstr <- svydesign(id = ~1, strata = ~region, weights = ~strwt, fpc = ~popsize,
                  data = agstrat)
dstr
## Stratified Independent Sampling design
## svydesign(id = ~1, strata = ~region, weights = ~strwt, fpc = ~popsize,
##     data = agstrat)
```

The syntax is similar to that for an SRS. The only difference is in the arguments to the *svydesign* function. The arguments used for stratification are as follows:

**id** As for an SRS, we use `id = ~1` to indicate that there is no clustering.

**strata** The *strata* argument gives the variable name(s) containing the stratification information (here, the stratification variable is *region*).

**weights** For stratified sampling, we use the weights associated with the selection probabilities in each stratum. This example has a stratified sample with proportional allocation, where the weights are almost identical for all strata. In samples with disproportionate stratification, however, the weights will vary across strata, and estimates calculated without weights will be biased.

**fpc** The *fpc* argument specifies the variable that contains information for calculating the finite population correction (fpc) in each stratum. The easiest way to do that is to create a new variable in the data frame that contains the population stratum sizes. Our code defines a variable *popsize_recode* that associates each stratum name with its population size; alternatively, the variable *popsize* could be created with the *merge* or *match* function or by assigning values separately to each region in a *for* loop.

If you omit the *fpc* argument (and still include the *weights* argument), the estimates of means and totals are the same, but standard error estimates are without the finite population correction.

All of the work specifying the design information is done in the *svydesign* function; after you have defined the design there, the *svymean* and *svytotal* functions are used exactly as in Chapter 2 for an SRS.

```
# calculate mean, SE and confidence interval
smean<-svymean(~acres92, dstr)
smean
##          mean    SE
## acres92 295561 16380
confint(smean, level=.95, df=degf(dstr)) # note that df = n-H = 300-4
##          2.5 %    97.5 %
## acres92 263325 327796.5
# calculate total, SE and CI
stotal<-svytotal(~acres92, dstr)
stotal
##           total        SE
## acres92 909736035 50417248
degf(dstr) # Show the degrees of freedom for the design
## [1] 296
# calculate confidence intervals using the degrees of freedom
confint(stotal, level=.95,df= degf(dstr))
##           2.5 %     97.5 %
```

```
## acres92 810514350 1008957721
```

The output is pretty self-explanatory. Note that 296 degrees of freedom $(\text{df} = n - H)$ are used for the confidence intervals. The df can also be found by applying function *degf* to the design object *dstr*, that is, `degf(dstr)`. If you want to calculate confidence intervals that are based on the normal distribution, simply omit the *df* argument in the *confint* function. If the sample has few observations, however, we need to specify the degrees of freedom and use the $t$ distribution to calculate confidence intervals.

**Weights and fpc arguments.** We supplied both *weights* and *fpc* arguments to the *svydesign* function in this example, but for a stratified random sample with no nonresponse, the *svydesign* function will calculate weights from the *fpc* information and the sample sizes in the data set. The design object *dstrfpc* in the following code results in the same statistics as the design object *dstr* (with the *weights* and *fpc* arguments) that we used earlier. Including the *weights* argument but omitting the *fpc* argument results in standard errors that are calculated without the fpc. (Do not omit both *weights* and *fpc*; then the *svydesign* function will assume all weights are equal.)

```
# Alternative design specifications
# Get same result if omit weights argument since weight = popsize/n_h
dstrfpc <- svydesign(id = ~1, strata = ~region, fpc = ~popsize, data = agstrat)
svymean(~acres92, dstrfpc)
##           mean     SE
## acres92 295561  16380
# If you include weights but not fpc, get SE without fpc factor
dstrwt <- svydesign(id = ~1, strata = ~region, weights = ~strwt, data = agstrat)
svymean(~acres92, dstrwt)
##           mean     SE
## acres92 295561  17241
```

**Calculating stratum means and variances.** Function *svyby* will calculate statistics and their standard errors for subgroups of the data. Here we use it to calculate the stratum means and totals. The first argument of *svyby* is the formula for the variable(s) for which statistics are desired, and the second argument (`by=`) is the variable that defines the groups. Then list the design object and the name of the function that calculates the statistics. Set `keep.var=TRUE` to display the standard errors for the statistics.

```
# calculate mean and se of acres92 by regions
svyby(~acres92, by=~region, dstr, svymean, keep.var = TRUE)
##     region   acres92        se
## NC      NC 300504.16  16107.59
## NE      NE  97629.81  18149.49
## S        S 211315.04  18925.35
## W        W 662295.51  93403.65
# calculate total and se of acres92 by regions
svyby(~acres92, ~region, dstr, svytotal, keep.var = TRUE)
##     region   acres92        se
## NC      NC 316731380  16977399
## NE      NE  21478558   3992889
## S        S 292037391  26154840
## W        W 279488706  39416342
```

If you want to check the calculations by formula, you can also calculate summary statistics directly for each stratum using the *tapply* function and then use the formulas from SDA to calculate the standard errors for each estimated stratum mean or total. The variances of the stratum means are calculated with the formula $(1 - n_h/N_h)s_h^2/n_h$, where $n_h$ and $N_h$ are the sample and population sizes, and $s_h^2$ is the sample variance within stratum $h$.

```
# formula calculations, using tapply
# variables sampsize and popsize were calculated earlier in the chapter
# calculate mean within each region
strmean<-tapply(agstrat$acres92,agstrat$region,mean)
strmean
##        NC        NE        S        W
## 300504.16   97629.81 211315.04 662295.51
# calculate variance within each region
strvar<-tapply(agstrat$acres92,agstrat$region,var)
strvar
##          NC          NE          S          W
## 29618183543   7647472708 53587487856 396185950266
# verify standard errors by direct formula
strse<- sqrt((1-sampsize/popsize)*strvar/sampsize)
# same standard errors as from svyby
strse
##
##        NC        NE        S        W
## 16107.59 18149.49 18925.35 93403.65
```

## 3.4   Estimating Proportions from a Stratified Random Sample

A proportion is a special case of a mean of a variable taking on values 1 and 0. As defined in Chapter 2, variable *lt200k* takes on value 1 if *acres92* < 200,000 and takes on value 0 if *acres92* ≥ 200,000. The mean of variable *lt200k* estimates the proportion of farms that have fewer than 200,000 acres. The total of variable *lt200k* estimates the number of farms that have fewer than 200,000 acres.

```
# Create variable lt200k
agstrat$lt200k <- rep(0,nrow(agstrat))
agstrat$lt200k[agstrat$acres92 < 200000] <- 1
# Rerun svydesign because the data set now has a new variable
dstr <- svydesign(id = ~1, strata = ~region, fpc = ~popsize,
  weights = ~strwt, data = agstrat)
# calculate proportion, SE and confidence interval
smeanp<-svymean(~lt200k, dstr)
smeanp
##          mean       SE
## lt200k 0.51391 0.0248
confint(smeanp, level=.95,df=degf(dstr))
##           2.5 %     97.5 %
## lt200k 0.4651188 0.5627107
# calculate total, SE and CI
stotalp<-svytotal(~lt200k, dstr)
stotalp
##         total       SE
## lt200k 1581.8 76.318
confint(stotalp, level=.95,df=degf(dstr))
##          2.5 %    97.5 %
## lt200k 1431.636 1732.024
```

You can also calculate proportions and totals of categorical variables by defining them to be factors, either by declaring the variable to be a factor variable in the data set or in the function call of *svymean*. Here we define variable *lt200kf* to be a factor variable in the data set.

```
# Create a factor variable lt200kf
agstrat$lt200kf <- factor(agstrat$lt200k)
# Rerun svydesign because the data set now has a new variable
dstr <- svydesign(id = ~1, strata = ~region, fpc = ~popsize,
                  weights = ~strwt, data = agstrat)
# calculate proportion, SE and confidence interval
smeanp2<-svymean(~lt200kf, dstr)
smeanp2
##               mean      SE
## lt200kf0 0.48609 0.0248
## lt200kf1 0.51391 0.0248
confint(smeanp2, level=.95,df=degf(dstr))
##                2.5 %     97.5 %
## lt200kf0 0.4372893 0.5348812
## lt200kf1 0.4651188 0.5627107
# calculate total, SE and CI
stotalp2<-svytotal(~lt200kf, dstr)
stotalp2
##           total      SE
## lt200kf0 1496.2 76.318
## lt200kf1 1581.8 76.318
confint(stotalp2, level=.95,df=degf(dstr))
##                2.5 %    97.5 %
## lt200kf0 1345.976 1646.364
## lt200kf1 1431.636 1732.024
```

Note that the *svytotal* function gives the estimated total for each category of variable *lt200kf*.

The **survey** package will also estimate asymmetric confidence intervals for survey data (Korn and Graubard, 1998), which may have more accurate coverage probabilities for proportions that are near 0 or 1 than the symmetric confidence intervals based on the normal approximation. This is done with the *svyciprop* function, choosing **method="beta"** to obtain a version of the Clopper-Pearson confidence interval (the function will also compute asymmetric confidence intervals using other methods). We illustrate with binary variable *lt200k*. Note that you need to list the formula as `~I(lt200k)` or `~I(lt200k==1)`.

```
# calculate proportion and confidence interval with svyciprop
svyciprop(~I(lt200k==1), dstr, method="beta")
##                        2.5% 97.5%
## I(lt200k == 1) 0.514 0.464  0.56
```

## 3.5    Additional Code for Exercises

Some of the exercises in Chapter 3 ask you to find an ANOVA table. Here's how to do that for *agstrat* using the *lm* function, which performs regression and analysis of variance. The first argument of *lm* is the formula for the regression model, of the form $y \sim x$. We specify *region* to be a factor so that the function will treat it as a categorical variable. (We'll see the function that conducts regression analyses for survey data in Chapter 4, and it will have a similar structure.)

```
myfit <- lm(acres92~factor(region), data=agstrat)
anova(myfit)
## Analysis of Variance Table
##
## Response: acres92
##                Df     Sum Sq    Mean Sq F value    Pr(>F)
## factor(region)  3 7.2976e+12 2.4325e+12   27.48 1.048e-15 ***
## Residuals     296 2.6202e+13 8.8521e+10
## ---
## Signif. codes:  0 '***' 0.001 '**' 0.01 '*' 0.05 '.' 0.1 ' ' 1
```

## 3.6    Summary, Tips, and Warnings

Table 3.2 lists the major functions used in this chapter to select or analyze data from a stratified random sample.

### Tips and Warnings

- When calculating optimal allocations, make sure that the variables containing the cost and variance information are in the same order as the variable(s) containing the stratum identifiers.

- Sort the population data set by the stratification variable(s) before calling the *strata* function to select a stratified sample.

- When calculating sampling weights for a stratified random sample, check that the sum of the sampling weights for each stratum equals the population size for that stratum.

- When analyzing data from a stratified random sample, first create the design object in the *svydesign* function, using the **strata=** argument. Then call the *svymean* and *svytotal* function with that design object.

- Functions *svymean*, *svytotal*, and *svyby* can be used to calculate statistics for two or more variables simultaneously. For example, svymean(~acres92 + acres87, dstr) will display statistics of both variables *acres92* and *acres87*.

**TABLE 3.2**
Functions used for Chapter 3.

| Function | Package | Usage |
|----------|---------|-------|
| order | base | Give indices for data sorted according to the specified variable |
| sample | base | Select a simple random sample with or without replacement |
| tapply | base | Apply a function to each group of values; groups are defined by the second argument |
| confint | stats | Calculate confidence intervals; add df for $t$ confidence interval |
| lm | stats | Fit a linear model to a data set (not using survey methods) |
| anova | stats | Calculate an analysis of variance table from a model object |
| boxplot | graphics | Draw boxplot of data (used to display strata in a stratified random sample) |
| strata | sampling | Select a stratified random sample |
| getdata | sampling | Extract the sampled units from the population |
| svydesign | survey | Specify the survey design; add stratum information for stratified random sample |
| degf | survey | Find degrees of freedom based on design information |
| svymean | survey | Calculate mean and standard error of mean (if the variable is numeric), or proportion in each category (if variable is categorical) |
| svytotal | survey | Calculate total and standard error of total |
| svyby | survey | Calculate survey statistics on subsets of a survey defined by factors |
| svyciprop | survey | Compute confidence intervals for proportions using various methods (if estimated proportions are close to 0 or 1, sometimes an asymmetric confidence interval is preferred to the symmetric confidence interval produced by *svymean*) |

# 4

## Ratio and Regression Estimation

Ratio and regression estimation both use auxiliary information to increase the precision of survey estimates. This chapter shows how to incorporate auxiliary information into survey data analyses using R. The code in this chapter is in file ch04.R on the book website.

## 4.1 Ratio Estimation

**Examples 4.2 and 4.3 of SDA.** The *svyratio* function in the **survey** package (Lumley, 2020) computes ratios from survey data. Let's see how it works for Examples 4.2 and 4.3 of SDA. As the correlation coefficient between variables *acres87* and *acres92* is 0.995806. *acres87* would be an excellent auxiliary variable for ratio estimation. The code and output to estimate the ratio $\bar{y}_\mathcal{U}/\bar{x}_\mathcal{U}$, where $\bar{y}_\mathcal{U}$ is the population mean of *acres92* and $\bar{x}_\mathcal{U}$ is the population mean of *acres87*, are given in the following.

```
data(agsrs)
n<-nrow(agsrs) #300
agsrs$sampwt <- rep(3078/n,n)
agdsrs <- svydesign(id = ~1, weights=~sampwt, fpc=rep(3078,300), data = agsrs)
agdsrs
## Independent Sampling design
## svydesign(id = ~1, weights = ~sampwt, fpc = rep(3078, 300), data = agsrs)
# correlatIon of acres87 and acres92
cor(agsrs$acres87,agsrs$acres92)
## [1] 0.995806
# estimate the ratio acres92/acres87
sratio<-svyratio(numerator = ~acres92, denominator = ~acres87,design = agdsrs)
sratio
## Ratio estimator: svyratio.survey.design2(numerator = ~acres92,
##     denominator = ~acres87, design = agdsrs)
## Ratios=
##           acres87
## acres92 0.9865652
## SEs=
##             acres87
## acres92 0.005750473
confint(sratio, df=degf(agdsrs))
##                     2.5 %    97.5 %
## acres92/acres87 0.9752487 0.9978818
```

The sample in *agsrs* is an SRS, so we specify the survey design object in *svydesign* exactly as we did in Chapter 2. The only new feature is the *svyratio* function, which calculates the ratio $\hat{B} = \bar{y}/\bar{x}$ and its standard error.

DOI: 10.1201/9781003228196-4

Now that we have estimated the ratio from the data, we can use the *predict* function to obtain the ratio estimates of the population mean and total of $y$. The population total of $x$ is $t_x = 964,470,625$ and the population mean of $x$ is $\bar{x}_\mathcal{U} = t_x/N$. Note that the value of $t_x$ came from the official U.S. Census of Agriculture statistics for 1987 (U.S. Bureau of the Census, 1995). This is greater than the sum of all $x$ values in data set *agpop* because some counties in the population have missing values for *acres87*, as we saw in Section 1.7.

```
# provide the population total of x
xpoptotal <- 964470625
# Ratio estimate of population total
predict(sratio,total=xpoptotal)
## $total
##             acres87
## acres92 951513191
##
## $se
##            acres87
## acres92 5546162
# Ratio estimate of population mean
predict(sratio,total=xpoptotal/3078)
## $total
##              acres87
## acres92 309133.6
##
## $se
##            acres87
## acres92 1801.872
```

Examples 4.2 and 4.3 of SDA also explore the scatterplot of *acres92* versus *acres87*. Because all of the weights are the same (=3078/300), we can use the base R function *plot* to display the data in Figure 4.1 (see Chapter 7 for how to draw scatterplots for samples with unequal weights).

We scale the $x$ and $y$ variables so that the plot shows millions of acres instead of acres, and specify the axis labels in the *xlab* and *ylab* arguments. The function *abline* draws the line through the origin with slope $\hat{B}$.

```
par(las=1) # make tick mark labels horizontal (optional)
plot(x=agsrs$acres87/1e6,y=agsrs$acres92/1e6,
  xlab="Millions of Acres Devoted to Farms (1987)",
  ylab = "Millions of Acres Devoted to Farms (1992)",
  main = "Acres Devoted to Farms in 1987 and 1992")
# draw line through origin with slope Bhat
abline(0,coef(sratio))
```

**Example 4.5 of SDA.** Variables of interest in this example are the number of woody seedlings in pig-protected areas under each of ten sampled oak trees in 1992 (*seed92*) and 1994 (*seed94*) on Santa Cruz Island, California. The code below draws the scatterplot (shown in Figure 4.4 of SDA and not reproduced here) of *seed94* versus *seed92*. It also calculates the correlation of the two variables.

```
#scatterplot of seed92 and seed94
data(santacruz)
plot(santacruz$seed92,santacruz$seed94,
     main="Number of seedlings in 1994 and 1992",
     xlab="Number of seedlings in 1992",ylab="Number of seedlings in 1994")
```

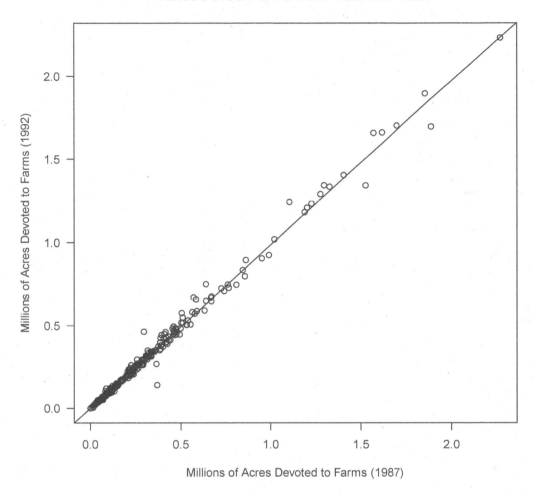

**FIGURE 4.1:** Scatterplot of acres devoted to farms in 1987 and 1992 (data *agsrs*).

```
cor(santacruz$seed92,santacruz$seed94)
## [1] 0.6106537
```

Now let's calculate the ratio of the number of seedlings in 1994 to the number of seedlings in 1992. Because the number of trees in the population is unknown, we define *sampwt* to be 1 for each observation, and we omit the *fpc* argument in the *svydesign* function (alternatively, one could set the population size to be a large number for the weights and fpc).

```
nrow(santacruz) #10
## [1] 10
santacruz$sampwt <- rep(1,nrow(santacruz))
design0405 <- svydesign(ids = ~1, weights = ~sampwt, data = santacruz)
design0405
## Independent Sampling design (with replacement)
## svydesign(ids = ~1, weights = ~sampwt, data = santacruz)
#Ratio estimation using number of seedlings of 1992 as auxiliary variable
```

```
sratio3<-svyratio(~seed94, ~seed92,design = design0405)
sratio3
## Ratio estimator: svyratio.survey.design2(~seed94, ~seed92, design = design0405)
## Ratios=
##           seed92
## seed94 0.2961165
## SEs=
##           seed92
## seed94 0.1152622
confint(sratio3, df=10-1)
##                    2.5 %      97.5 %
## seed94/seed92 0.03537532 0.5568577
```

## 4.2   Regression Estimation

**Example 4.7 of SDA.** Function *svyglm* calculates regression coefficients and regression estimators from survey data. It is the survey analog of the R function *glm*, which fits generalized linear models.

```
data(deadtrees)
head(deadtrees)
##   photo field
## 1    10    15
## 2    12    14
## 3     7     9
## 4    13    14
## 5    13     8
## 6     6     5
nrow(deadtrees) # 25
## [1] 25
# Fit with survey regression
dtree<- svydesign(id = ~1, weight=rep(4,25), fpc=rep(100,25), data = deadtrees)
myfit1 <- svyglm(field~photo, design=dtree)
summary(myfit1) # displays regression coefficients
##
## Call:
## svyglm(formula = field ~ photo, design = dtree)
##
## Survey design:
## svydesign(id = ~1, weight = rep(4, 25), fpc = rep(100, 25), data = deadtrees)
##
## Coefficients:
##             Estimate Std. Error t value Pr(>|t|)
## (Intercept)   5.0593     1.3930   3.632   0.0014 **
## photo         0.6133     0.1259   4.870 6.44e-05 ***
## ---
## Signif. codes:  0 '***' 0.001 '**' 0.01 '*' 0.05 '.' 0.1 ' ' 1
##
## (Dispersion parameter for gaussian family taken to be 5.548341)
##
## Number of Fisher Scoring iterations: 2
confint(myfit1,df=23) # df = 25-2
```

```
##                2.5 %    97.5 %
## (Intercept) 2.3291420 7.7894421
## photo       0.3664593 0.8600894
# Regression estimate of population mean field trees
newdata <- data.frame(photo=11.3)
predict(myfit1, newdata)
##     link    SE
## 1 11.989 0.418
confint(predict(myfit1, newdata),df=23)
##      2.5 %   97.5 %
## 1 11.12455 12.85404
# Estimate total field tree, add population size in total= argument
newdata2 <- data.frame(photo=1130)
predict(myfit1, newdata2, total=100)
##     link     SE
## 1 1198.9 41.802
confint(predict(myfit1, newdata2,total=100),df=23)
##       2.5 %   97.5 %
## 1 1112.455 1285.404
```

The regression estimation uses the following functions:

**svydesign** As before, use *svydesign* to describe the design of the survey, which, in this case, is an SRS with sample size $n = 25$ and population size $N = 100$. Each observation has a weight of $100/25 = 4$.

**svyglm** The function `svyglm(field~photo, design=dtree)` tells which variables to analyze in the regression statement. The dependent ($y$) variable is before the ~ sign, and the independent ($x$) variables follow it. In this example, the dependent variable is *field* and there is one independent variable, *photo*. The *design* argument tells the name of the survey design object (here, *dtree*) to use in calculations.

**summary** The *summary* function gives the estimates of the regression coefficients, their associated standard errors, and the $t$ statistic and $p$-value for testing whether each regression parameter equals 0 (the standard errors and tests will be discussed in Chapter 11).

**confint** As before, the *confint* function requests confidence limits for the regression parameters. You can specify the degrees of freedom with the *df* argument if desired: here the df equal the sample size minus 2: $25 - 2 = 23$. If you omit the *df* argument, the function uses the normal distribution to produce confidence intervals.

**predict** The *predict* function allows you to obtain estimates for predicted values from the estimated regression equation. For regression estimation of the mean, we first define a new data frame with `photo = 11.3` because we want to calculate the predicted value of the regression function at $\bar{x}_{\mathcal{U}} = 11.3$. The statement `predict(myfit1, newdata)` gives $\hat{B}_0 * (1) + \hat{B}_1 * (11.3)$. Regression estimation of the total multiplies each of these by the population size $N$ (here, $N = 100$) by including the argument `total=100` in the *predict* function (called as `predict(myfit1, newdata2, total=100)`), which estimates $\hat{B}_0 * (100) + \hat{B}_1 * (1130)$, where $t_x = 1130$.

Note that the output gives slightly different standard errors and confidence intervals for the regression estimates of the mean and total than SDA because the R functions use a slightly different (although asymptotically equivalent) formula to calculate the standard error. See Section 11.6 of SDA for a discussion of the two variance estimates used.

## 4.3   Domain Estimation

A domain is a subset of the population for which estimates are desired. Because estimated domain means and totals are ratio estimates, they can be calculated with the *svyratio* function. It is usually easier, however, to compute them using *subset* or *svyby*.

The procedure to calculate estimates for domains is essentially the same as that to calculate estimates for the full sample, but you need to redefine the design for the domain with the *subset* function so that standard errors are calculated correctly. Type:

```
newdesign<-subset(original_design, domain)
```

**Example 4.8 of SDA.** The following code uses the *subset* function to request design information for each level of the variable *farmcat*, which is defined to equal "large" when *farms92* $\geq 600$ and "small" otherwise.

```
agsrsnew<-agsrs #copy agsrs as agsrsnew, since we want to create a new column
# we calculated sampwt in the first code in this chapter
# define new variable farmcat
agsrsnew$farmcat<-rep("large",n)
agsrsnew$farmcat[agsrsnew$farms92 < 600] <- "small"
head(agsrsnew)
##               county state acres92 acres87 acres82 farms92 farms87 farms82
## 1      COFFEE COUNTY    AL  175209  179311  194509     760     842     944
## 2     COLBERT COUNTY    AL  138135  145104  161360     488     563     686
## 3       LAMAR COUNTY    AL   56102   59861   72334     299     362     447
## 4     MARENGO COUNTY    AL  199117  220526  231207     434     471     622
## 5      MARION COUNTY    AL   89228  105586  113618     566     658     748
## 6   TUSCALOOSA COUNTY   AL   96194  120542  134616     436     521     650
##    largef92 largef87 largef82 smallf92 smallf87 smallf82 region sampwt farmcat
## 1        29       28       21       57       47       66      S  10.26   large
## 2        37       41       42       12       44       47      S  10.26   small
## 3         4        4        3       16       20       30      S  10.26   small
## 4        48       66       62       14       11       28      S  10.26   small
## 5         7        9        9       11       23       27      S  10.26   small
## 6        20       17       23       18       32       29      S  10.26   small
dsrsnew <- svydesign(id = ~1, weights=~sampwt, fpc=rep(3078,300), data=agsrsnew)
# domain estimation for large farmcat with subset statement
dsub1<-subset(dsrsnew,farmcat=='large')   # design info for domain large farmcat
smean1<-svymean(~acres92,design=dsub1)
smean1
##            mean    SE
## acres92 316566 21553
df1<-sum(agsrsnew$farmcat=='large')-1 #calculate domain df if desired
df1
## [1] 128
confint(smean1, level=.95,df=df1) # CI
##           2.5 %   97.5 %
## acres92 273918.9 359212.4
stotal1<-svytotal(~acres92,design=dsub1)
stotal1
##            total        SE
## acres92 418987302 38938277
confint(stotal1, level=.95,df=df1)
```

```
##              2.5 %     97.5 %
## acres92 341941269 496033335
# domain estimation for small farmcat
dsub2<-subset(dsrsnew,farmcat=='small')  # design info for domain small farmcat
smean2<-svymean(~acres92,design=dsub2)
smean2
##           mean      SE
## acres92 283814 28852
df2<-sum(agsrsnew$farmcat=='small')-1 #calculate domain df if desired
confint(smean2, level=.95,df=df2) #CI
##              2.5 %     97.5 %
## acres92 226858.9 340768.5
stotal2<-svytotal(~acres92,design=dsub2)
stotal2
##             total       SE
## acres92 497939808 55919525
confint(stotal2, level=.95,df=df2)
##              2.5 %     97.5 %
## acres92 387553732 608325884
```

You can also calculate statistics for all domains defined by a factor variable at the same time, using the *svyby* function. Here, we estimate the population total and mean for both domains defined by *factor(farmcat)*. The first argument of *svyby* contains the variable(s) to analyze, and the second argument is the factor variable that defines the domains. The last argument gives the name of the function that is to be applied to each group in the *by* argument.

```
bothtot<-svyby(~acres92,by=~factor(farmcat),design=dsrsnew,svytotal)
bothtot
##       factor(farmcat)   acres92        se
## large           large 418987302 38938277
## small           small 497939808 55919525
confint(bothtot,level=.95)
##           2.5 %     97.5 %
## large 342669682 495304922
## small 388339553 607540062
bothmeans<-svyby(~acres92,by=~factor(farmcat),design=dsrsnew,svymean)
bothmeans
##       factor(farmcat)   acres92       se
## large           large 316565.7 21553.21
## small           small 283813.7 28852.24
confint(bothmeans,level=.95)
##           2.5 %     97.5 %
## large 274322.1 358809.2
## small 227264.4 340363.1
```

Note that confidence intervals here are slightly smaller than those given from the calculations with the *subset* function and in Example 4.8 of SDA. Because we did not specify the df in the *confint* function, it uses a normal distribution to calculate the intervals; the previous code, using the *subset* function, calculated the confidence intervals using a $t$ distribution having $n_d - 1$ df, where $n_d$ is the sample size of domain $d$.

**Warning.** In SRSs, you can calculate domain means and their standard errors by first forming a new, subsetted data set that consists of the observations in the domain and then calculating statistics on the subsetted data set. In complex surveys, however, that method can result in incorrect standard errors (see Section 11.3 of SDA). To obtain correct statistics for domains, first define the survey design object using the function *svydesign* with the entire data set. Then use the function *subset* or *svyby* with the survey design object to obtain correct inferences for domains.

## 4.4   Poststratification

**Example 4.9 of SDA.** The *postStratify* function computes poststratification weights and uses them to estimate population means and totals, along with their standard errors (discussed in Chapter 11 of SDA). Let's poststratify the SRS in *agsrs* by variable *region*.

```
data(agsrs)
dsrs <- svydesign(id = ~1, weights=rep(3078/300,300), fpc=rep(3078,300),
                  data = agsrs)
# Create a data frame that gives the population totals for the poststrata
pop.region <- data.frame(region=c("NC","NE","S","W"), Freq=c(1054,220,1382,422))
# create design information with poststratification
dsrsp<-postStratify(dsrs, ~region, pop.region)
summary(dsrsp)
## Independent Sampling design
## postStratify(dsrs, ~region, pop.region)
## Probabilities:
##    Min. 1st Qu.  Median    Mean 3rd Qu.    Max.
## 0.09242 0.09407 0.09407 0.09771 0.10152 0.10909
## Population size (PSUs): 3078
## Data variables:
##  [1] "county"   "state"    "acres92"  "acres87"  "acres82"  "farms92"
##  [7] "farms87"  "farms82"  "largef92" "largef87" "largef82" "smallf92"
## [13] "smallf87" "smallf82" "region"
1/unique(dsrsp$prob)  # See the poststratified weight for each region
## [1] 10.630769 10.820513  9.850467  9.166667
svymean(~acres92, dsrsp)
##           mean    SE
## acres92 299778 17513
svytotal(~acres92, dsrsp)
##             total        SE
## acres92 922717031 53906392
```

The only new feature here is the *postStratify* function:

```
postStratify(design=dsrs, strata=region, population=pop.region)
```

The *postStratify* function tells R to construct poststratification weights, using poststrata in the variable *region*. The third argument is the name of the data frame (here, *pop.region*) that gives the population totals for the poststrata. The poststratified estimates of the population mean and total of *acres92*, when calculated with poststratified design object *dsrsp*, are reported together with standard errors when the *svymean* or *svytotal* function is called.

Note that the standard errors reported by the `survey` package differ slightly from those in Example 4.9 of SDA because a slightly different (although asymptotically equivalent) estimator for the variance is used (see Section 11.6 of SDA).

## 4.5 Ratio Estimation with Stratified Sampling

The *svyratio* function will compute either separate or combined ratio estimates. The default is combined ratio estimation, which calculates the ratio $\hat{\bar{y}}/\hat{\bar{x}}$, where $\hat{\bar{y}}$ is the estimate of the mean of $y$ using the stratified design and $\hat{\bar{x}}$ is the estimated mean of $x$. All we need to do is to include the stratification information in the design structure formed by the *svydesign* function.

**Combined ratio estimator.** The following shows how to compute the ratio of *acres92* to *acres87* and the ratio estimator of the total for the stratified sample in *agstrat*, using the combined ratio estimator.

```
data(agstrat)
popsize_recode <- c('NC' = 1054, 'NE' = 220, 'S' = 1382, 'W' = 422)
agstrat$popsize <- popsize_recode[agstrat$region]
# input design information for agstrat
dstr <- svydesign(id = ~1, strata = ~region, fpc = ~popsize, weight = ~strwt,
                  data = agstrat)
# now compute the combined estimator of the ratio
combined<-svyratio(~ acres92,~acres87,design = dstr)
combined
## Ratio estimator: svyratio.survey.design2(~acres92, ~acres87, design = dstr)
## Ratios=
##            acres87
## acres92 0.9899971
## SEs=
##              acres87
## acres92 0.006187757
# we can get the combined ratio estimator of the population total
# with the predict function
predict(combined,total=964470625)
## $total
##            acres87
## acres92 954823130
##
## $se
##            acres87
## acres92 5967910
```

**Separate ratio estimator.** You can calculate ratios separately for each stratum by including separate=TRUE in the *svyratio* function.

```
separate<-svyratio(~acres92,~acres87,design = dstr,separate=TRUE)
separate
## Stratified ratio estimate: svyratio.survey.design2(~acres92, ~acres87,
##     design = dstr, separate = TRUE)
## Ratio estimator: Stratum == "NC"
## Ratios=
##            acres87
## acres92 0.9750666
## SEs=
##              acres87
## acres92 0.005483458
## Ratio estimator: Stratum == "NE"
```

```
## Ratios=
##              acres87
## acres92 0.8956073
## SEs=
##               acres87
## acres92 0.008853011
## Ratio estimator: Stratum == "S"
## Ratios=
##               acres87
## acres92 0.9935483
## SEs=
##                acres87
## acres92 0.01418835
## Ratio estimator: Stratum == "W"
## Ratios=
##            acres87
## acres92 1.011974
## SEs=
##                acres87
## acres92 0.01169809
#  Define the stratum totals for acres87 as a list:
stratum.xtotals <- list(NC=350474227,NE=22033421,S=280631939,W=311331038)
predict(separate,stratum.xtotals)
## $total
##             acres87
## acres92 955349448
##
## $se
##          acres87
## acres92 5731438
```

## 4.6   Model-Based Ratio and Regression Estimation

This section is optional and need only be read if covering Section 4.6 of SDA.

**Example 4.11 of SDA.** A model-based analysis of data from an SRS uses the same techniques taught in an introductory statistics class. Since the model-based analysis does not make use of the sampling weights, the *lm* or *glm* functions, which fit linear and generalized linear models for non-survey data, are used to fit the regression models and obtain the residuals. Here we use the *lm* function.

The format for fitting a regression model with *lm* is similar to *svyglm*, but with one important difference: the weights mean different things in the two functions. In the *svyglm* function, `weights=` tells how many population units are represented by each sample unit. In the *lm* function, the weight variable contains relative weights for a weighted least squares fit.

The model used is $Y_i = \beta x_i + \varepsilon_i$, with $V(\varepsilon_i) = \sigma^2 x_i$. The model has variance proportional to $x_i$, so obtaining the best linear unbiased estimates under this model would use a weight value proportional to the reciprocal of the variances. This is specified by defining the variable $recacr87 = 1/acres87$ when $acres87 > 0$ and $recacr87 = $ NA when $acres87 = 0$ (to avoid division by zero).

```
data(agsrs)
# define weights to use for weighted least squares analysis
agsrs$recacr87<-agsrs$acres87
agsrs$recacr87[agsrs$acres87!=0] <- 1/agsrs$acres87[agsrs$acres87!=0]
agsrs$recacr87[agsrs$acres87==0] <- NA
# fit weighted least squares model without intercept
fit<-lm(acres92~acres87-1,weights=recacr87,data=agsrs)
summary(fit)
##
## Call:
## lm(formula = acres92 ~ acres87 - 1, data = agsrs, weights = recacr87)
##
## Weighted Residuals:
##    Min   1Q Median    3Q    Max
## -369.9  -22.2   -5.8   10.8  311.7
##
## Coefficients:
##         Estimate Std. Error t value Pr(>|t|)
## acres87 0.986565   0.004844   203.7   <2e-16 ***
## ---
## Signif. codes:  0 '***' 0.001 '**' 0.01 '*' 0.05 '.' 0.1 ' ' 1
##
## Residual standard error: 46.1 on 298 degrees of freedom
##   (1 observation deleted due to missingness)
## Multiple R-squared:  0.9929,Adjusted R-squared:  0.9928
## F-statistic: 4.149e+04 on 1 and 298 DF,  p-value: < 2.2e-16
anova(fit)
## Analysis of Variance Table
##
## Response: acres92
##            Df   Sum Sq  Mean Sq F value    Pr(>F)
## acres87     1 88168461 88168461   41487 < 2.2e-16 ***
## Residuals 298   633307     2125
## ---
## Signif. codes:  0 '***' 0.001 '**' 0.01 '*' 0.05 '.' 0.1 ' ' 1
# find predicted value at population total for x
newdata3 <- data.frame(acres87=964470625)
predict(fit, newdata3, se.fit=TRUE)
## $fit
##         1
## 951513191
##
## $se.fit
## [1] 4671509
##
## $df
## [1] 298
##
## $residual.scale
## [1] 46.0998
```

The *weights* argument in *lm* specifies that a weighted least squares analysis is performed with weights *recacr87*, minimizing the weighted sum of squares $\sum_{i \in \mathcal{S}} (y_i - \beta x_i)^2 / x_i$. The "$-1$" in lm(acres92~acres87-1) tells that the model is to be fit without an intercept. The *summary* function displays the regression coefficient $\hat{\beta} = 0.986565$ and the *anova* function displays the ANOVA table. The model is fit to the 299 observations that have *acres87* > 0.

The *predict* function requests the predicted value from the regression model when *acres87* takes on the value $t_x = 964{,}470{,}625$, giving $\hat{t}_{yM} = \hat{\beta}t_x = 951{,}513{,}191$. The standard error, without the fpc, is $\sigma t_x / \sqrt{\sum_{i \in \mathcal{S}} x_i} = 4{,}671{,}509$.

Note that the sum of squares for error in the ANOVA table, 633,307, is the sum of squares of the weighted residuals, so the mean squared error in the ANOVA table gives $\hat{\sigma}^2 = 2125.19$.

The residuals produced by *lm* are $e_i = y_i - \hat{y}_i$. For a ratio model, the weighted residuals $e_{iw} = e_i / \sqrt{x_i}$ should be plotted instead of $e_i$, because if the model variance structure is correct, the $e_{iw}$ should all have approximately equal variances and a plot of $e_{iw}$ versus the predicted values or $x_i$ will show no patterns.

```
# plot weighted residual versus predicted values
wresid<-fit$residuals*sqrt(fit$weights)
par(las=1)
plot(fit$fitted.values, wresid,
    main="Plot of weighted residuals versus predicted values",
    xlab="Predicted value from regression model",
    ylab="Weighted residuals")
```

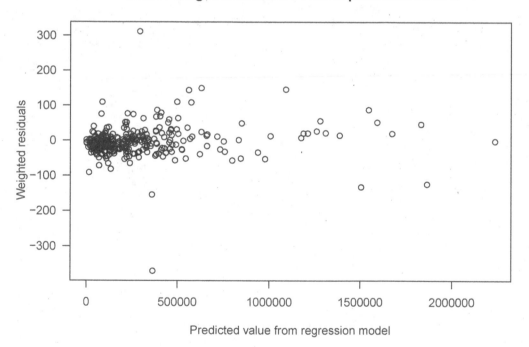

**FIGURE 4.2:** Plot of weighted residuals versus fitted values.

Figure 4.2 shows a couple of potential outliers, but no other indications that the model is inappropriate.

**Example 4.12 of SDA.** The *lm* function is also used to fit a regression model and to obtain residuals for the dead tree data from Example 4.7 of SDA.

```
data(deadtrees)
# Fit with lm
fit2 <- lm(field~photo, data=deadtrees)
summary(fit2)
##
## Call:
## lm(formula = field ~ photo, data = deadtrees)
##
## Residuals:
##     Min      1Q  Median      3Q     Max
## -5.0319 -1.8053  0.1947  1.4212  3.8080
##
## Coefficients:
##             Estimate Std. Error t value Pr(>|t|)
## (Intercept)   5.0593     1.7635   2.869 0.008676 **
## photo         0.6133     0.1601   3.832 0.000854 ***
## ---
## Signif. codes:  0 '***' 0.001 '**' 0.01 '*' 0.05 '.' 0.1 ' ' 1
##
## Residual standard error: 2.406 on 23 degrees of freedom
## Multiple R-squared:  0.3896,Adjusted R-squared:  0.3631
## F-statistic: 14.68 on 1 and 23 DF,  p-value: 0.0008538
# Estimate mean field trees
newdata <- data.frame(photo=11.3)
predict(fit2, newdata,se.fit=TRUE)
## $fit
##        1
## 11.98929
##
## $se.fit
## [1] 0.4941007
##
## $df
## [1] 23
##
## $residual.scale
## [1] 2.406153
```

Because the regression model is fit under the assumption that $V(\varepsilon_i) = \sigma^2$ for all observations, no *weights* argument is used in *lm*. The *predict* function gives the regression estimate of the population mean $\hat{\beta}_0 + \hat{\beta}_1\bar{x}_{\mathcal{U}} = 11.9893$, and its standard error (without fpc) of 0.494. These are the values calculated in Example 4.12 of SDA. Typing summary(fit2) gives the regression coefficients, their standard errors, and other information about the fit.

We have applied both *lm* and *svyglm* to analyze the tree data from Example 4.7. Table 4.1 compares the estimates and standard errors from the two functions. All the point estimates are the same, but the standard errors from *svyglm* differ from those calculated by *lm*; Sections 4.6 and 11.4 of SDA discuss why that occurs.

**TABLE 4.1**

Comparison of the estimates and standard errors for model `field~photo` from the *lm* and *svyglm* functions.

|  | Intercept | | Slope | | Predicted Value, $x = 11.3$ | |
|---|---|---|---|---|---|---|
|  | lm | svyglm | lm | svyglm | lm | svyglm |
| Estimate | 5.0593 | 5.0593 | 0.6133 | 0.6133 | 11.989 | 11.989 |
| SE | 1.7635 | 1.3930 | 0.1601 | 0.1259 | 0.494 | 0.418 |

Requesting `plot(fit2)` produces a collection of residual and diagnostic plots from the *lm* model object. Figure 4.3 displays the plot of the residuals versus *photo* (the $x$ variable), which shows no pattern.

```
# plot residuals versus predicted values
plot(deadtrees$photo, fit2$residuals,
    main="Plot of residuals versus photo values",
    xlab="Photo values (x variable)",
    ylab="Residuals")
```

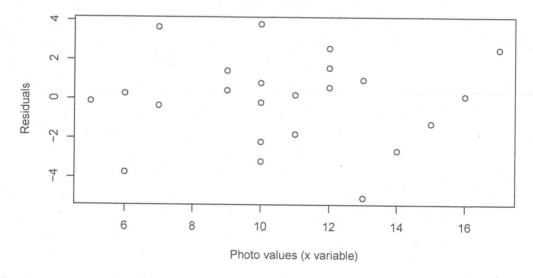

**FIGURE 4.3:** Plot of residuals versus $x$ variable.

## 4.7  Summary, Tips, and Warnings

Table 4.2 lists the major functions used in this chapter to compute ratio and regression estimates, calculate statistics for domains, and create poststratification weights.

To calculate a ratio or ratio estimator for an SRS or stratified sample, use the *svyratio* function from the **survey** package. The function *svyglm* fits regression models to survey data.

**Tips and Warnings**

- Draw a scatterplot of your data when fitting a ratio or regression model, so you can see whether ratio or regression estimation is likely to improve efficiency.

- For domain estimation, first define a design object for the entire sample with the *svydesign* function. Then use the *subset* function to define a domain of interest, or calculate statistics for all domains with the *svyby* function. It may be tempting to calculate statistics for a subset of the sample by creating a data set containing only that subset, but doing that can result in incorrect standard errors for domain statistics.

- Poststratification can be done using the *postStratify* function.

**TABLE 4.2**

Functions used for Chapter 4.

| Function | Package | Usage |
|---|---|---|
| subset | base | Work with a subset of a vector, matrix, or data frame |
| confint | stats | Calculate confidence intervals |
| cor | stats | Calculate the correlation of vectors (not using survey methods) |
| lm | stats | Fit a linear model to a data set (not using survey methods) |
| anova | stats | Compute an analysis of variance table from a model object |
| predict | stats | Obtain predicted values from a model object |
| plot | graphics | Draw a scatterplot of data |
| abline | graphics | Add a straight line to a plot |
| svydesign | survey | Specify the survey design |
| svymean | survey | Calculate mean and standard error of mean |
| svyratio | survey | Calculate a ratio or ratio estimate from survey data |
| svyglm | survey | Fit a regression model to survey data. The coefficients may then be used to calculate regression estimates |
| svytotal | survey | Calculate total and standard error of total |
| svyby | survey | Calculate statistics for subsets of a survey defined by a factor variable |
| postStratify | survey | Adjust the sampling weights using poststratification |

# 5

# Cluster Sampling with Equal Probabilities

This chapter shows how to use R to compute estimates from one- and two-stage cluster samples when an SRS is selected at each stage. Chapter 6 will tell how to select a one-stage or two-stage cluster sample with equal or unequal probabilities—the syntax is similar for both, and deferring the sample selection examples to Chapter 6 allows us to look at the general case. The code in this chapter is in file `ch05.R` on the book website.

## 5.1  Estimates from One-Stage Cluster Samples

**Example 5.2 of SDA.** The following code and output produces estimates of the population mean and total for the GPA data, using functions from the **survey** package (Lumley, 2020). Variable *suite* identifies the clusters in the data, and variable *wt* is the sampling weight for the persons in the sample, defined as $100/5 = 20$ for every person.

```
data(gpa)
# define one-stage cluster design
# note that id is suite instead of individual student as we take an SRS of suites
dgpa<-svydesign(id=~suite,weights=~wt,fpc=~rep(100,20),data=gpa)
dgpa
## 1 - level Cluster Sampling design
## With (5) clusters.
## svydesign(id = ~suite, weights = ~wt, fpc = ~rep(100, 20), data = gpa)
# estimate mean and se
gpamean<-svymean(~gpa,dgpa)
gpamean
##       mean      SE
## gpa 2.826 0.1637
degf(dgpa)
## [1] 4
# n=5, t-approximation is suggested for CI
confint(gpamean,level=.95,df=4) # use t-approximation
##       2.5 %   97.5 %
## gpa 2.371593 3.280407
# confint(gpamean,level=.95) # uses normal approximation, if desired (for large n)
# estimate total and se (if desired)
gpatotal<-svytotal(~gpa,dgpa)
gpatotal
##       total      SE
## gpa 1130.4 65.466
confint(gpatotal,level=.95,df=4)
##       2.5 %   97.5 %
## gpa 948.6374 1312.163
```

The following features of the code and output deal with the cluster sampling:

- The `id=~suite` argument in *svydesign* tells all functions that use the design object that variable *suite* contains the primary sampling unit (psu) identifiers. If you omit the psu variable and instead use `id=~`, the data will be (incorrectly) analyzed as an SRS.

- The `weights=~wt` argument says that variable *wt* contains the sampling weights for the observation units (here, the students). If `weights=` is omitted but `fpc=` is supplied, selection probabilities are calculated from the population sizes, assuming an SRS of psus. That results in the same weights for this survey, but we recommend including the `weights=` argument as routine practice because most surveys have some type of adjustment that causes the final weights to differ from the sampling weights. (There is one exception, and that is for the without-replacement variance calculations discussed in Section 5.2.)

- The `fpc=~rep(100,20)` argument indicates that the total number of psus in the population is 100. When the *fpc* argument is included, functions use the finite population correction (fpc) when calculating variances. Omit `fpc=` if you do not want an fpc (but then make sure you include the `weights=` argument).

- Typing `dgpa` shows that this is a "1 - level Cluster Sampling design With (5) clusters". The *svydesign* function recognizes this as a one-stage design because one clustering variable is included in the *id* argument. This is the only indication in the output that the clustering was used in the analysis. Otherwise, the form of the statistics output is the same as for simple random or stratified sampling. Always check that the number of clusters in the design object equals the number of psus in your sample.

- The degrees of freedom (df), from function *degf*, equals the number of psus minus 1.

- The *svymean* and *svytotal* functions produce standard errors (SEs) for estimated population means and totals that account for the clustering in the design.

You can verify the calculations using the formulas given in Section 5.2 of SDA if desired.

```
# you can also calculate SEs by direct formula
suitesum<-tapply(gpa$gpa,gpa$suite,sum)  #sum gpa for each suite
# variability comes from among the suites
st2<-var(suitesum)
st2
## [1] 2.25568
# SE of t-hat, formula (5.3) of SDA
vthat <-100^2*(1-5/100)*st2/5
sqrt(vthat)
## [1] 65.46596
# SE of ybar, formula (5.6) of SDA
sqrt(vthat)/(4*100)
## [1] 0.1636649
```

The variability *st2* is coming from the suite totals, and the fpc $(1 - 5/100)$ is applied to calculate the variance of $\hat{t}$. The SE of $\hat{t}$ is 65.46596, which is the same as calculated by *svytotal*.

The procedure for calculating estimates from one-stage cluster samples is exactly the same when the psus have unequal sizes.

**Example 5.6 of SDA.** Data set *algebra* has 12 classes (clusters) with unequal sizes selected from 187 classes. The syntax for analyzing the data with unequal-sized clusters is exactly the same as for Example 5.2. Again, note that we use the number of clusters minus 1 (= 11) as the df.

```
data(algebra)
algebra$sampwt<-rep(187/12,299)
# define one-stage cluster design
dalg<-svydesign(id=~class,weights=~sampwt,fpc=~rep(187,299), data=algebra)
dalg
## 1 - level Cluster Sampling design
## With (12) clusters.
## svydesign(id = ~class, weights = ~sampwt, fpc = ~rep(187, 299),
##     data = algebra)
# estimate mean and se
svymean(~score,dalg)
##         mean      SE
## score 62.569 1.4916
# n=12, t-distribution is suggested for CI
degf(dalg)
## [1] 11
confint(svymean(~score,dalg),level=.95,df=11) #use t-approximation
##         2.5 %  97.5 %
## score 59.28562 65.8515
# estimate total and se if desired
svytotal(~score,dalg)
##         total      SE
## score 291533 19893
confint(svytotal(~score,dalg),level=.95,df=11)
##         2.5 %    97.5 %
## score 247749.4 335316.6
```

## 5.2   Estimates from Multi-Stage Cluster Samples

Calculating estimates from a multi-stage cluster sample is similar. The clustering structure is specified for the design object in the *svydesign* function, and then all functions called with that design object account for the clustering in the variance calculations.

There are several ways to estimate variances using the **survey** package. Let's start with Example 5.8 of SDA, where we calculate the with-replacement variance, and then discuss the issues involved for calculating variances for without-replacement samples using the schools data in Example 5.7 of SDA.

**Example 5.8 of SDA.** The *coots* data come from Arnold's (1991) work on egg size and volume of American coot eggs in Minnedosa, Manitoba, with a sample of 184 clutches (nests of eggs). Variable *csize* gives the number of eggs in the clutches. Two eggs (secondary sampling unit, ssu) are randomly selected from each clutch (psu). Since we do not have information on the total number of psus $N$, we use the relative weights *relwt* defined by *csize/2* to calculate the mean volume of eggs and its standard error.

```
data(coots)
# Want to estimate the mean egg volume
nrow(coots) #368
## [1] 368
```

```
coots$ssu<-rep(1:2,184) # index of ssu
coots$relwt<-coots$csize/2
head(coots)
##   clutch csize length breadth   volume tmt ssu relwt
## 1      1    13  44.30   31.10 3.7957569   1   1   6.5
## 2      1    13  45.90   32.70 3.9328497   1   2   6.5
## 3      2    13  49.20   34.40 4.2156036   1   1   6.5
## 4      2    13  48.70   32.70 4.1727621   1   2   6.5
## 5      3     6  51.05   34.25 0.9317646   0   1   3.0
## 6      3     6  49.35   34.40 0.9007362   0   2   3.0
dcoots<-svydesign(id=~clutch+ssu,weights=~relwt,data=coots)
dcoots
## 2 - level Cluster Sampling design (with replacement)
## With (184, 368) clusters.
## svydesign(id = ~clutch + ssu, weights = ~relwt, data = coots)
svymean(~volume,dcoots)  #ratio estimator
##          mean     SE
## volume 2.4908 0.061
confint(svymean(~volume,dcoots),level=.95,df=183)
##         2.5 %   97.5 %
## volume 2.370423 2.611134
# now only include psu information, results are the same
dcoots2<-svydesign(id=~clutch,weights=~relwt,data=coots)
dcoots2
## 1 - level Cluster Sampling design (with replacement)
## With (184) clusters.
## svydesign(id = ~clutch, weights = ~relwt, data = coots)
svymean(~volume,dcoots2)
##          mean     SE
## volume 2.4908 0.061
```

In *svydesign*, the two stages of the cluster sampling are given as id=~clutch+ssu. The formula lists the psus first and then the ssus. When the with-replacement variance is calculated, however, as is done here, you need only specify the psus—the point and variance estimates are the same whether you specify just the psus or you specify all stages of sampling. The *weight* argument must be included for this design because the weights are unequal.

Note that the confidence interval uses a *t* critical value with 183 df (number of psus minus 1). Also note that *svydesign* does not contain the *fpc* argument. This is because the total number of clutches in the population, $N$, is unknown. As a result, the *svymean* does not use an fpc when calculating estimates. In general, we recommend omitting the *fpc* argument for multi-stage cluster sampling even when $N$ is known, and the remainder of this section discusses this issue.

**Variance estimation for without-replacement two-stage cluster samples.** Here are two options for estimating the variance of estimated means and totals in without-replacement two-stage sampling, where an SRS is selected at each stage.

**Option 1. Calculate the with-replacement variance (recommended).** As shown in Sections 5.3 and 6.6 of SDA, the estimated variability among estimated psu totals, $s_t^2$, also includes variability from the subsequent stages of sampling. If you estimate the with-replacement variance (at the psu level), the variance estimator incorporates *all* the variability from subsequent stages of sampling. The expected value of the with-replacement variance estimator is larger than the true variance of the without-replacement sample, but the difference is small if the sampling fraction at the psu level, $n/N$, is small.

Chapter 6 of SDA outlines additional benefits of ignoring the fpcs when the psus are selected with unequal probabilities.

To estimate the with-replacement variance for a multi-stage cluster sample, simply call the *svydesign* function as:

```
svydesign(id=~psuid,weights=~weightvariable,data=dataset)
```

where the *id* formula consists only of the variable giving the psu membership. Do not include the *fpc* argument when calling *svydesign*. Then *svytotal*, *svymean*, and other functions will calculate the with-replacement variance.

**Option 2. Calculate the without-replacement variance.** When an SRS or stratified random sample is taken at all stages of sampling, you can specify all stages of sampling in the *svydesign* function and calculate without-replacement variances. For a two-stage sample, call the *svydesign* function as:

```
svydesign(id=~psuid+ssuid,fpc=~psufpc+ssufpc,data=dataset).
```

The *id* formula gives the variable identifying the psu membership followed by the variable identifying the ssu membership. The *fpc* formula has *psufpc*, the variable giving the population number of psus (= `rep(N,nrow(dataset))`), followed by *ssufpc*, the variable giving the values of the psu size for each observation (= $M_i$ for ssus in psu $i$).

No *weights* argument is included. When the *weights* argument is omitted, it is assumed that an SRS is taken at both stages, and the inclusion probabilities are calculated from the population sizes given in *fpc* and the sample sizes in the data set. Thus, the weights are assumed to be $(NM_i)/(nm_i)$, where the values of $n$ and $m_i$ are counted from the data set and the values of $N$ and $M_i$ are given in the *fpc* arguments.

If there are more than two stages of sampling, and an exactly unbiased estimate of the variance is desired, you need to include terms for all stages of sampling in the *id* and *fpc* arguments of the *svydesign* function. If a survey has three stages, the without-replacement variance estimate requires knowledge of the psu membership, population size, and sample size; the ssu membership, population size, and sample size; and the tertiary sampling unit membership, population size, and sample size.

It can be complicated to keep track of all this information. In addition, calculations are done under the assumption that the final weights are the same as the sampling weights (computed as the inverse of the inclusion probabilities)—that is, there are no nonresponse adjustments or other modifications of the sampling weights.

**Example 5.7 of SDA: With-replacement variance.** Let's look at the with- and without-replacement variance calculations for the *schools* data. The following code calculates the with-replacement variance. Note that only the psu-level clustering is specified in the *id* argument and that the vector of student-level weights is provided. We can also estimate the proportion and the total number of students having *mathlevel*=2 by treating *mathlevel* as a factor variable.

```
data(schools)
head(schools)
##   schoolid gender math reading mathlevel readlevel  Mi finalwt
## 1        9      F   42      42         2         2 163  61.125
## 2        9      F   29      30         1         1 163  61.125
## 3        9      M   31      25         1         1 163  61.125
## 4        9      F   22      33         1         2 163  61.125
## 5        9      M   35      36         1         2 163  61.125
```

```
## 6          9      F   30      17          1        1 163  61.125
# calculate with-replacement variance; no fpc argument
# include psu variable in id; include weights
dschools<-svydesign(id=~schoolid,weights=~finalwt,data=schools)
# dschools tells you this is treated as a with-replacement sample
dschools
## 1 - level Cluster Sampling design (with replacement)
## With (10) clusters.
## svydesign(id = ~schoolid, weights = ~finalwt, data = schools)
mathmean<-svymean(~math,dschools).
mathmean
##        mean     SE
## math 33.123 1.7599
degf(dschools)
## [1] 9
# use t distribution for confidence intervals because there are only 10 psus
confint(mathmean,df=degf(dschools))
##          2.5 %  97.5 %
## math 29.14179 37.1041
# estimate proportion and total number of students with mathlevel=2
svymean(~factor(mathlevel),dschools)
##                       mean      SE
## factor(mathlevel)1 0.71231 0.0542
## factor(mathlevel)2 0.28769 0.0542
svytotal(~factor(mathlevel),dschools)
##                     total      SE
## factor(mathlevel)1 12303.4 2244.14
## factor(mathlevel)2  4969.1  676.26
```

**Example 5.7 of SDA: Without-replacement variance.** The *svymean* function will calculate without-replacement variances when simple or stratified random sampling is used at each stage. (As of this writing, it does not do so for all designs and thus will not compute the without-replacement variance for most of the unequal-probability samples that are discussed in Chapter 6.) To use it with the *schools* data, put both stages of clustering in the *id* argument and put both the psu and the ssu population sizes in the *fpc* argument. Do not include the *weights* argument.

```
# create a variable giving each student an id number
schools$studentid<-1:(nrow(schools))
# calculate without-replacement variance
# specify both stages of the sample in the id argument
# give both sets of population sizes in the fpc argument
# do not include the weight argument
dschoolwor<-svydesign(id=~schoolid+studentid,fpc=~rep(75,nrow(schools))+Mi,
                      data=schools)
dschoolwor
## 2 - level Cluster Sampling design
## With (10, 200) clusters.
## svydesign(id = ~schoolid + studentid, fpc = ~rep(75, nrow(schools)) +
##     Mi, data = schools)
mathmeanwor<-svymean(~math,dschoolwor)
mathmeanwor
##        mean     SE
## math 33.123 1.6605
confint(mathmeanwor,df=degf(dschoolwor))
##          2.5 %  97.5 %
```

```
## math 29.36667 36.87923
# estimate proportion and total number of students with mathlevel=2
svymean(~factor(mathlevel),dschoolwor)
##                       mean      SE
## factor(mathlevel)1 0.71231 0.0516
## factor(mathlevel)2 0.28769 0.0516
svytotal(~factor(mathlevel),dschoolwor)
##                      total      SE
## factor(mathlevel)1 12303.4 2097.83
## factor(mathlevel)2  4969.1  657.69
```

In the *schools* data, variable *Mi* gives the population number of students in each school. This information must be available in the data set to be able to calculate the without-replacement variance. The design object *dschoolwor* repeats that this is a "2-level Cluster Sampling Design" with 10 psus and 200 ssus.

Even with the relatively large sampling fractions in this example, the with- and without-replacement standard errors are similar. For variable *math*, the with-replacement standard error is 1.76, and the without-replacement standard error is 1.66.

In general, we recommend calculating the with-replacement variance (omitting the *fpc* argument) for multi-stage cluster sampling. It produces a variance estimate whose expectation is slightly larger than the true variance, but if $n/N$ is small, the difference is negligible. If forced to choose between a standard error that is slightly too large and one that is too small, we usually prefer the former because a too-small standard error leads to claiming that estimates are more precise than they really are.

The most important thing to keep in mind for computing standard errors for cluster samples is that ssus in the same psu are usually more homogeneous than randomly selected ssus from the population. Thus, the essential feature for calculating standard errors is to capture that homogeneity by including the `id=~psuid` argument in *svydesign*. The issue of "to fpc or not to fpc" is minor compared with the effect of clustering.

## 5.3  Model-Based Design and Analysis for Cluster Samples

We often use models when designing a cluster sample, as shown in Section 5.4 of SDA. Data from a previous survey or pilot sample may be used to estimate the optimal subsampling or psu size. This often involves estimating the value of $R^2$ or $R_a^2$, which can be obtained from an Analysis of Variance (ANOVA) table.

**Example 5.12 of SDA.** The following shows how to derive an ANOVA table for the schools data.

```
# run lm with schoolid as a factor
fit5.12<-lm(math~factor(schoolid), data=schools)
# print ANOVA table
anova(fit5.12)
## Analysis of Variance Table
##
## Response: math
##                  Df  Sum Sq Mean Sq F value    Pr(>F)
## factor(schoolid)  9  7018.5  779.83  7.5834 1.785e-09 ***
```

```
## Residuals          190 19538.4  102.83
## ---
## Signif. codes:  0 '***' 0.001 '**' 0.01 '*' 0.05 '.' 0.1 ' ' 1
# extract the value of R-squared and adjusted R-squared
summary(fit5.12)$r.squared
## [1] 0.264281
summary(fit5.12)$adj.r.squared
## [1] 0.2294312
```

**Example 5.14 of SDA.** This example employs a random effects model, in which the school means are assumed to be normally distributed random variables with mean $\mu$. In packages **nlme** (Pinheiro et al., 2021) and **lme4** (Bates et al., 2015, 2020), the *lme* (short for linear mixed effects) and *lmer* functions, respectively, calculate estimates from random effects models.

We use function *lme* for this example. For the one-way random effects model, the only fixed effect is the mean, so the *fixed* formula is `fixed=math~1`. Random effects are specified in the *random* argument so *factor(schoolid)*, the factor variable describing the psu membership, is placed behind the vertical bar in the *random* argument.

```
library(nlme)
fit5.14 <- lme(fixed=math~1,random=~1|factor(schoolid),data=schools)
summary(fit5.14)
## Linear mixed-effects model fit by REML
##   Data: schools
##        AIC      BIC    logLik
##   1516.259 1526.139 -755.1295
##
## Random effects:
##  Formula: ~1 | factor(schoolid)
##         (Intercept) Residual
## StdDev:    5.818064 10.14069
##
## Fixed effects:  math ~ 1
##              Value Std.Error  DF  t-value p-value
## (Intercept) 34.66  1.974628 190 17.55267       0
##
## Standardized Within-Group Residuals:
##         Min          Q1         Med          Q3         Max
## -2.26713655 -0.74262324 -0.09451607  0.79142521  2.18576500
##
## Number of Observations: 200
## Number of Groups: 10
# extract the variance components
VarCorr(fit5.14)
## factor(schoolid) = pdLogChol(1)
##             Variance  StdDev
## (Intercept)  33.84987  5.818064
## Residual    102.83368 10.140694
```

A model-based analysis predicts the values of observations that are not observed in the data. For this data set, the unobserved values are the students who are not measured in the sampled schools, as well as the unsampled schools in the population. The estimated mean 34.66 from the output under "`Fixed effects: math ~ 1`" does not account for the population sizes of the schools, and gives a different estimate than in Example 5.7 of SDA.

The residuals and predicted values from the model can be obtained by requesting `resid(fit5.14)` and `predict(fit5.14)`. You can also type `plot(fit5.14)` to obtain a plot of standardized residuals versus fitted values.

## 5.4 Additional Code for Exercises

**Exercise 5.40 of SDA.** The exercise uses the function *intervals_ex40*, available from the book website and R package `SDAResources` (Lu and Lohr, 2021). To run the function, load the `SDAResources` package and type

```
intervals_ex40(groupcorr=0, numintervals=100, groupsize=5, sampgroups=10,
               popgroups=5000, mu=0, sigma=1)
```

using the desired values for the arguments. The function call given above uses the default values of the arguments and will give the same results as running `intervals_ex40()`.

The arguments of function *intervals_ex40* are given in Table 5.1.

**TABLE 5.1**
Arguments of the *intervals_ex40* function.

| Argument | Description |
|---|---|
| groupcorr | Desired intraclass correlation coefficient, must be between 0 and 1 (default is 0). |
| numintervals | Number of confidence intervals to be generated (default is 100). |
| groupsize | Number of observations, $M$, in each population cluster (default is 5). |
| sampgroups | Number of clusters to be sampled (default is 10). |
| popgroups | Number of clusters in population (default is 5000). This should be set to be at least 200 times as large as the value of *sampgroups* so that the fpc is negligible. |
| mu | Population mean (default is 0). |
| sigma | Population standard deviation (default is 1). |

For the exercise, you are asked to generate 100 intervals of 50 observations each, taken in 10 clusters of size 5. This uses the default values of all arguments except for ICC. When running the function, you only need to specify the arguments that differ from the default values, so that you can generate 100 intervals with ICC = 0.3 by running the function with argument `groupcorr=0.3`.

```
set.seed(9231)
# generate intervals for cluster sample with groupcorr = 0.3
intervals_ex40(groupcorr = 0.3) # leave other parameters unchanged
##   Number_of_intervals      SRS_cover_prob    Cluster_cover_prob
##          100.0000000           0.8000000             0.9500000
##     SRS_mean_CI_width Cluster_mean_CI_width
##            0.5556272            0.9111856
##       Replicate mu sample_mean      srs_lci     srs_uci in_srs_ci SRS_CI_width
## [1,]          1  0 -0.07311322 -0.35806539  0.21183894         1    0.5699043
## [2,]          2  0  0.25038517 -0.05127437  0.55204471         1    0.6033191
## [3,]          3  0  0.03499401 -0.27232646  0.34231448         1    0.6146409
## [4,]          4  0 -0.18948478 -0.45038074  0.07141119         1    0.5217919
```

```
## [5,]          5  0  0.14058713 -0.08585203 0.36702629        1     0.4528783
##         clus_lci  clus_uci  in_clu_ci clus_CI_width
## [1,] -0.6526766 0.5064502          1    1.1591268
## [2,] -0.2716028 0.7723731          1    1.0439759
## [3,] -0.4941334 0.5641214          1    1.0582548
## [4,] -0.5814274 0.2024579          1    0.7838853
## [5,] -0.2344528 0.5156271          1    0.7500799
```

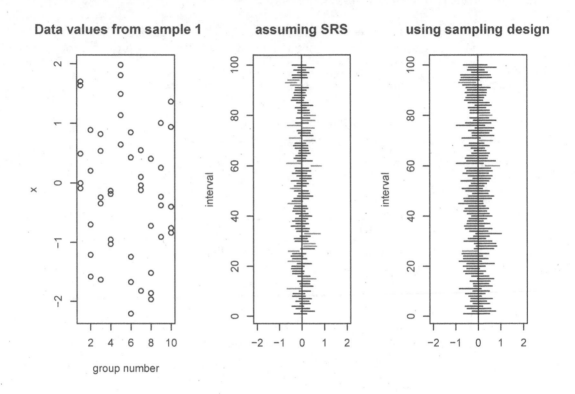

**FIGURE 5.1:** Interval estimates created assuming SRS and using clustering formulas.

We initialized the random number seed so that we could reproduce the intervals later, but if you are repeating the exercise, you may want to let the computer generate your starting seed (otherwise, you may get the same set of samples each time).

- Function *intervals_ex40* calculates two sets of interval estimates: a set that uses SRS formulas and hence has coverage probability (proportion of intervals that include the true population mean 0) SRS_cover_prob of 0.80 that is less than 0.95, and a second set that calculates the correct confidence intervals using the formulas for one-stage cluster sampling with coverage probability Cluster_cover_prob of 0.95.

- It also prints the average width of the interval estimates for the two methods: SRS_mean _CI_width of 0.5556272, which is less than Cluster_mean_CI_width of 0.9111856.

- The first five replicates and their summary statistics are printed, where srs_lci is the lower limit, and srs_uci is the upper limit from the SRS estimate. Similarly, clus_lci and clus_uci are the lower and upper confidence limits from the estimate calculated using the clustering. If desired, the function can be modified so that information from all replicates is stored in a data set.

- Three graphs are produced, similar to those in Figure 5.1. The first graph shows a scatterplot of the last simulated sample, the second graph shows the interval estimates produced for each sample if analyzed as an SRS, and the third shows the interval estimates produced for each sample when analyzed as a cluster sample. When the graph is produced in color, intervals that include the true value of the population mean are black, and those that do not include the true value are red.

The estimated coverage probability for each procedure is the proportion of intervals that include the true population mean. In Figure 5.1, the estimated coverage probability of the procedure that (incorrectly) treats the data as an SRS is 0.80—substantially smaller than the nominal 0.95 probability of a confidence interval.

## 5.5   Summary, Tips, and Warnings

Table 5.2 lists the main R functions used in this chapter. We have seen most of these before; the main difference is how the survey design is specified in the *svydesign* function to indicate the clustering.

**TABLE 5.2**

Functions used for Chapter 5.

| Function | Package | Usage |
|----------|---------|-------|
| tapply | base | Apply a function to each group of values; groups are defined by the second argument |
| confint | stats | Calculate confidence intervals, add df for $t$ confidence interval |
| lm | stats | Fit a linear model to a data set (not using survey methods) |
| lme | nlme | Fit a random-effects or mixed-effects model to a data set (not using survey methods) |
| svydesign | survey | Specify the survey design |
| svymean | survey | Calculate mean and standard error of mean |
| svytotal | survey | Calculate total and standard error of total |
| intervals_ex40 | SDAResources | Show differences between (incorrect) SRS formulas and (correct) cluster formulas applied to cluster samples |

In the **survey** package, clusters are identified in the **id=** argument of the *svydesign* function. The general form of the *svydesign* function for a one-stage cluster sample (without stratification) is:

```
svydesign(id=~psuvar,weights=~weightvar,fpc=rep(N,nobs),data=dataset)
```

where *psuvar* is the name of the variable in *dataset* containing the psu identifiers. The variable *weightvar* contains the weights for the observation units in the data. If an fpc is desired (and often it is not), $N$ is the number of psus in the population, and *nobs* is the number of psus in the sample.

For multi-stage cluster sampling (again without stratification), the following form of the *svydesign* function will calculate point estimates with the weights in *weightvar* and with-replacement standard errors (note the absence of the fpc argument):

`svydesign(id=~psuvar,weights=~weightvar,data=dataset).`

After specifying the survey design, the *svymean* and *svytotal* functions are used exactly as in other chapters. The only difference for cluster sampling is that you must list the cluster variable(s) in the *svydesign* function.

**Tips and Warnings**

- Use the *id* argument to specify the clustering, and check that the number of clusters listed when you print the survey design object equals the number of psus in your sample.

- When calculating estimates for one-stage cluster samples, or for two-stage cluster samples using with-replacement variance estimates, include the *weights* argument when specifying the survey design. The weight variable should contain the final weights at the observation level. Check that the sum of the weights approximately equals the number of observation units in the population.

- In general, we recommend calculating with-replacement variances, but the `survey` package functions will also calculate without-replacement variances for the designs discussed in this chapter, where an SRS is taken at each stage of sampling. If you calculate the without-replacement variances for a two-stage cluster design, it is useful to check these by also calculating the with-replacement variance (they should be close).

# 6

## Sampling with Unequal Probabilities

In this chapter, we discuss how to select a sample with equal or unequal probabilities, and how to compute estimates from an unequal probability sample. The code is in file `ch06.R` on the book website.

Let's start with sample selection. Section 6.1 shows how to select a one-stage cluster sample with equal or unequal probabilities, and Section 6.2 presents two methods for selecting a two-stage sample. We'll look at code for computing estimates for unequal-probability samples in Section 6.3.

## 6.1 Selecting a Sample with Unequal Probabilities

This section shows how to select a sample of primary sampling units (psus) with unequal probabilities with the *sample* function and with functions from the `sampling` package (Tillé and Matei, 2021). Subsampling all secondary sampling units (ssus) in the selected psus will give a one-stage cluster sample.

### 6.1.1 Sampling with Replacement

**Example 6.2 of SDA.** In Chapters 2 through 5, units, whether observation units or clusters, were selected with equal probabilities. In Example 6.2, 5 classes are sampled from 15 classes with probability proportional to size (pps) and with replacement.

Section 2.1 showed how to use the *sample* function to select a simple random sample, with or without replacement. It can also be used to select a with-replacement sample with unequal probabilities by including the optional *prob* argument. Call the function as

```
sample(1:N,n,replace=TRUE,prob=probvar)
```

where $N$ is the number of psus in the population, $n$ is the desired sample size of psus, and *probvar* is a vector of length $N$ that gives the size measures or selection probabilities for the psus. In this example, psus are classes, and *class_size* gives the number of students in the class.

```
data(classes)
classes[1:2,]
##    class class_size
## 1      1         44
## 2      2         33
N<-nrow(classes)
set.seed(78065)
# select 5 classes with probability proportional to class size and with replacement
sample_units<-sample(1:N,5,replace=TRUE,prob=classes$class_size)
```

```
sample_units
## [1]  5 14  6 14  6
mysample<-classes[sample_units,]
mysample
##        class class_size
## 5         5         76
## 14       14        100
## 6         6         63
## 14.1     14        100
## 6.1       6         63
# calculate ExpectedHits and sampling weights
mysample$ExpectedHits<-5*mysample$class_size/sum(classes$class_size)
mysample$SamplingWeight<-1/mysample$ExpectedHits
mysample$psuid<-row.names(mysample)
mysample
##        class class_size ExpectedHits SamplingWeight psuid
## 5         5         76    0.5873261       1.702632     5
## 14       14        100    0.7727975       1.294000    14
## 6         6         63    0.4868624       2.053968     6
## 14.1     14        100    0.7727975       1.294000  14.1
## 6.1       6         63    0.4868624       2.053968   6.1
# check sum of sampling weights
sum(mysample$SamplingWeight)
## [1] 8.398568
```

- Note that classes 6 and 14 both appear twice in the sample. When collecting data in one stage, each student within classes 6 and 14 must be included twice. Otherwise, estimates will be biased. If collecting data in two stages, you would take two independent subsamples from class 6 and two independent subsamples from class 14.

  When analyzing the data, make sure you use different psu names for the multiple instances of psus that appear more than once. For this example, you might want to use `row.names(mysample)` as the psu identifier, since it gives a unique name to each sampled psu.

- After selecting the sample, we need to calculate the sampling weights. The *ExpectedHits* variable gives $n\psi_i$, where $n$ is the sample size and $\psi_i$ is the draw-by-draw selection probability that is proportional to *class_size* (note that the values of $\psi_i$ sum to 1). This is the number of times we expect the unit to be in the sample. For example, class 5 has *ExpectedHits* = 5*76/647 = 0.5873261.

  Then, the *SamplingWeight* is $1/ExpectedHits = 1/(n\psi_i)$ for with-replacement sampling.

- The sum of the sampling weights for the sample is an unbiased estimator of $N$, and for large samples it should be close to $N$. For this example, however, the weight sum has a large standard error because the sample size is so small (see Exercise 6.45 of SDA).

## 6.1.2   Sampling without Replacement

There are several functions in the `sampling` package that will select unequal-probability samples without replacement. When the list of units in the sampling frame is in random order, systematic sampling is likely to produce a sample that behaves like an SRS without replacement. The *cluster* function in the `sampling` package can select a pps sample using systematic sampling.

The following code shows how the *cluster* function is called.

```
set.seed(330582)
cluster(data=classes, clustername=c("class"), size=5, method="systematic",
        pik=classes$class_size,description=TRUE)
## Number of selected clusters: 5
## Number of units in the population and number of selected units: 15 5
##   class ID_unit      Prob
## 1     1       1 0.3400309
## 2     5       5 0.5873261
## 3     8       8 0.3400309
## 4    11      11 0.3554869
## 5    14      14 0.7727975
```

The following arguments of the *cluster* function are used:

- `pik` is the vector of inclusion probabilities (or a vector of relative unit sizes that can be used to compute the probabilities). In this example, *pik* is *class_size*, the auxiliary variable that gives the size of each class.

- `size=5` requests a sample of 5 units.

- `method="systematic"` describes the method used to select the sample. The *cluster* function can also be called with methods `"srswor"` (simple random sampling without replacement), `"srswr"` (simple random sampling with replacement), or `"poisson"` (Poisson sampling). The *pik* argument is not needed with methods srswor and srswr.

- `description=TRUE` asks the function to print the number of population and sampled units.

The *cluster* function creates variable *Prob* that includes the final inclusion probabilities for the units in the sample. You can compute sampling weights as $1/Prob$.

Other functions are also available for selecting unequal-probability samples. Table 6.1 lists sample selection functions in the **sampling** package that correspond to methods discussed in Chapters 5 and 6 of SDA. Tillé (2006) describes these and additional methods in the **sampling** package (the functions that select unequal-probability samples have names that begin with *UP*) for selecting samples. Another resource is the **pps** package (Gambino, 2021), which contains several functions for selecting unequal-probability samples.

You may want to consider writing your own function or using a different software package if none of the methods implemented in R meet your sample selection needs. The SURVEYSELECT procedure in SAS software provides additional options for selecting without-replacement probability samples, including the Hanurav–Vijayan method (Hanurav, 1967; Vijayan, 1968); see SAS Institute Inc. (2021) and Lohr (2022) for details.

## 6.2 Selecting a Two-Stage Cluster Sample

There are several ways to select a two-stage cluster sample in R. The *mstage* function from the **sampling** package will select both stages at once for simple random, systematic, or Poisson sampling. Alternatively, you can select the units at each stage separately: First select the psus, then select a sample of ssus from the selected psus.

**TABLE 6.1**
Some functions for selecting a probability sample in the `sampling` package.

| Function | Description |
|---|---|
| UPbrewer(pik) | Select an unequal-probability sample without replacement containing 2 psus per stratum using Brewer's (1963, 1975) method. The *pik* argument contains the desired inclusion probabilities $\pi_i$. Note that there is no argument for the sample size. For all of the UP sample selection methods, the sample size is assumed to be implicit in the *pik* vector because $\sum_{k=1}^{N} \pi_k = n$. The function `inclusionprobabilities(size,n)` will compute *pik* from a vector *size* of positive numbers and desired sample size *n*. |
| UPpoisson(pik) | Select a sample (of variable size) using Poisson sampling. The *pik* argument contains the desired inclusion probabilities $\pi_k$ for each unit, and these should be between 0 and 1. |
| UPsystematic(pik) | Select an unequal-probability sample via systematic sampling. |
| UPsampford(pik) | Select an unequal-probability sample using Sampford's (1967) method, an extension of Brewer's method that allows drawing more than 2 psus per stratum. |
| srswor(n,N) | Select an SRS of size *n* without replacement from a population of size *N*. |
| srswr(n,N) | Select an SRS of size *n* with replacement from a population of size *N*. |

**Example 6.11 of SDA: Selection with *mstage* function.** In Chapter 1, we expanded data `classes` to a long format that includes student information. Let's redo that here.

```
# create data frame classeslong
data(classes)
classeslong<-classes[rep(1:nrow(classes),times=classes$class_size),]
classeslong$studentid <- sequence(classes$class_size)
nrow(classeslong)
## [1] 647
table(classeslong$class) # check class sizes
##
##   1   2   3   4   5   6   7   8   9  10  11  12  13  14  15
##  44  33  26  22  76  63  20  44  54  34  46  24  46 100  15
head(classeslong)
##     class class_size studentid
## 1       1         44         1
## 1.1     1         44         2
## 1.2     1         44         3
## 1.3     1         44         4
## 1.4     1         44         5
## 1.5     1         44         6
```

We now use the *mstage* function to select a pps systematic sample of 5 classes and take an SRS without replacement of 4 students from each class. Each class in the psu sample can therefore be considered as a stratum for sample selection purposes, and an independent sample of size 4 is taken from each stratum. We call the function as

```
mstage(classeslong,stage=list("cluster","stratified"),
            varnames=list("class","studentid"),
            size=numberselect, method=list("systematic","srswor"),pik=prob)
```

In the function *mstage*, sampling specifications for the different stages are given in list objects. Lists in R allow you to combine structures of different type; here, the lists consist of vectors that have different lengths. If there are four stages of sampling, each list will have four components, each in the order of the stages of sampling. This example is for two-stage sampling, so each list has two components.

- The two stages are denoted by `stage=list("cluster","stratified")`, and the corresponding list naming the stratification or clustering variables at the stages is `varnames=list("class","studentid")`.

  For this example, a cluster sample of psus, identified by *class*, is desired for the first stage of sampling. After the psus are selected, the frame for the second stage of sampling consists of the listing of ssus for the sample of 5 psus. An SRS is selected from each of those 5 psus, so the method used is `"stratified"`.

- The desired sample sizes are given in the *size* argument as a list containing two levels. Here, we set *size* equal to the list `numberselect<-list(5,rep(4,5))`. We want to select a sample of 5 psus and then a subsample of 4 ssus from each sampled psu.

- The sampling methods at the two stages, systematic and simple random sampling respectively, are denoted by `method=list("systematic","srswor")`.

- The selection probabilities are given in `pik=prob`, where

  ```
  prob<-list(classes$class_size/647,4/classeslong$class_size)
  or
  prob<-list(classes$class_size/647)
  # srswor is with probability of 4/classeslong$class_size by default
  # since ssu sample size of 4 is supplied in numberselect argument
  ```

As always, you can set the seed to any integer you like; this allows you to re-create the sample later. Note that the sample in this book has different psus than the sample in SDA, which was selected using SAS software (SAS Institute Inc., 2021).

```
# select a two-stage cluster sample, psu: class, ssu: studentid
# number of psus selected: n = 5 (pps systematic)
# number of students selected: m_i = 4 (srs without replacement)
# problist<-list(classes$class_size/647) # same results as next command
problist<-list(classes$class_size/647,4/classeslong$class_size) #selection prob
problist[[1]]  # extract the first object in the list. This is pps, size M_i/M
##  [1] 0.06800618 0.05100464 0.04018547 0.03400309 0.11746522 0.09737249
##  [7] 0.03091190 0.06800618 0.08346213 0.05255023 0.07109737 0.03709428
## [13] 0.07109737 0.15455951 0.02318393
problist[[2]][1:5] # first 5 values in second object in list
## [1] 0.09090909 0.09090909 0.09090909 0.09090909 0.09090909
# number of psus and ssus
n<-5
numberselect<-list(n,rep(4,n))
numberselect
## [[1]]
## [1] 5
##
## [[2]]
```

```
## [1] 4 4 4 4 4
# two-stage sampling
set.seed(75745)
tempid<-mstage(classeslong,stage=list("cluster","stratified"),
               varnames=list("class","studentid"),
               size=numberselect, method=list("systematic","srswor"),pik=problist)
```

The output *tempid* contains two objects

```
sample1<-getdata(classeslong,tempid)[[1]]
sample2<-getdata(classeslong,tempid)[[2]].
```

Here, *sample1* contains the classes selected at stage 1 along with the selection probabilities *Prob_ 1 _stage*, and *sample2* contains the selected ssus and the second-stage sampling probabilities, as well as the final selection probabilities *Prob* (which equals the product of *Prob_ 1 _stage* and *Prob_ 2 _stage*). We only need *sample2* but also show *sample1* so you can see the first-stage probabilities.

```
# get data
sample1<-getdata(classeslong,tempid)[[1]]
# sample 1 contains the ssus of the 5 psus chosen at the first stage
# Prob_ 1 _stage has the first-stage selection probabilities
head(sample1)
##        class_size studentid class ID_unit Prob_ 1 _stage
## 4.21          22        22     4     125      0.1700155
## 4.20          22        21     4     124      0.1700155
## 4.6           22         7     4     110      0.1700155
## 4             22         1     4     104      0.1700155
## 4.7           22         8     4     111      0.1700155
## 4.8           22         9     4     112      0.1700155
nrow(sample1)
## [1] 285
table(sample1$class) # lists the psus selected in the first stage
##
##   4   6   9  13  14
##  22  63  54  46 100
sample2<-getdata(classeslong,tempid)[[2]]
# sample 2 contains the final sample
# Prob_ 2 _stage has the second-stage selection probabilities
# Prob has the final selection probabilities
head(sample2)
##        class class_size studentid ID_unit Prob_ 2 _stage       Prob
## 4.21       4         22        22     125     0.18181818 0.0309119
## 4.7        4         22         8     111     0.18181818 0.0309119
## 4.5        4         22         6     109     0.18181818 0.0309119
## 4.19       4         22        20     123     0.18181818 0.0309119
## 6.48       6         63        49     250     0.06349206 0.0309119
## 6.53       6         63        54     255     0.06349206 0.0309119
nrow(sample2) # sample of 20 ssus altogether
## [1] 20
table(sample2$class) # 4 ssus selected from each psu
##
##  4  6  9 13 14
##  4  4  4  4  4
# calculate final weight = 1/Prob
sample2$finalweight<-1/sample2$Prob
```

```
# check that sum of final sampling weights equals population size
sum(sample2$finalweight)
## [1] 647
sample2[,c(1,2,3,6,7)] # print variables from final sample
##         class class_size studentid      Prob finalweight
## 4.21      4        22        22 0.0309119       32.35
## 4.7       4        22         8 0.0309119       32.35
## 4.5       4        22         6 0.0309119       32.35
## 4.19      4        22        20 0.0309119       32.35
## 6.48      6        63        49 0.0309119       32.35
## 6.53      6        63        54 0.0309119       32.35
## 6.23      6        63        24 0.0309119       32.35
## 6.33      6        63        34 0.0309119       32.35
## 9.50      9        54        51 0.0309119       32.35
## 9.29      9        54        30 0.0309119       32.35
## 9.31      9        54        32 0.0309119       32.35
## 9.36      9        54        37 0.0309119       32.35
## 13.10    13        46        11 0.0309119       32.35
## 13       13        46         1 0.0309119       32.35
## 13.45    13        46        46 0.0309119       32.35
## 13.39    13        46        40 0.0309119       32.35
## 14.4     14       100         5 0.0309119       32.35
## 14.78    14       100        79 0.0309119       32.35
## 14.98    14       100        99 0.0309119       32.35
## 14.63    14       100        64 0.0309119       32.35
```

The final sampling weight *finalweight* is the reciprocal of final selection probability. The psus were selected with probabilities $5M_i/647$ and the ssus for psu $i$ were selected with probability $4/M_i$, so the final weight is $[647/(5M_i)](M_i/4) = 647/20 = 32.35$ for each ssu in the sample.

**Example 6.11 of SDA: Selection in two steps.** Another option, if you want to use a method other than systematic sampling to select the psus with unequal probabilities, is to select the first-stage units and the second-stage units in separate steps. This is often more convenient for populations where the psu sizes $M_i$ are known only for units in the sample. For example, if nursing homes are psus, you may have to find out the value of $M_i$ directly from each home and thus would know these values only after the first-stage sample is selected.

Let's select a sample from data set *classes* in two stages. In this example we use function *UPsampford* (see Table 6.1) instead of systematic sampling to select 5 classes (psus) at the first stage of sample selection. We call

```
UPsampford(pik)
```

where *pik* is a vector of length $N$ containing the desired inclusion probabilities. The function has no argument for the sample size; it is assumed that sum(pik)=n. The function *inclusionprobabilities* will compute *pik*, having sum $n$, from the desired sample size and a vector of positive numbers that gives the relative sizes of the units.

```
# select a cluster sample in two stages, psu: class, ssu: studentid
# number of psu selected n =5 (Sampford's method)
# first, convert the measure of size to a vector of probabilities
classes$stage1prob<-inclusionprobabilities(classes$class_size,5)
sum(classes$stage1prob)  # inclusion probabilities sum to n
## [1] 5
```

```
# select the psus
set.seed(29385739)
stage1.units<-UPsampford(classes$stage1prob)
stage1.sample<-getdata(classes,stage1.units)
stage1.sample
##    ID_unit class class_size stage1prob
## 1        1     1         44  0.3400309
## 3        3     3         26  0.2009274
## 7        7     7         20  0.1545595
## 13      13    13         46  0.3554869
## 14      14    14        100  0.7727975
```

The data frame *stage1.sample* contains the psus (classes) for the sample. Variable *stage1prob* contains the first-stage selection probabilities that we computed from the *class_size* variable. Now we can use the function to select the second-stage units. Since we have already formed the data frame *stage1.sample* that consists only of the sampled psus, the *strata* function provides a convenient way to select an SRS from each sampled psu.

To draw an SRS of students from each sampled psu, we first create data frame *stage1.long* with a data record for each student in the sampled classes, and then take an SRS of 4 students from each class.

```
# first-stage units are in stage1.sample
# now select the second-stage units (students)
# convert the psus in the sample to long format and assign student ids
npsu<-nrow(stage1.sample)
stage1.long<-stage1.sample[rep(1:npsu,times=stage1.sample$class_size),]
stage1.long$studentid<-sequence(stage1.sample$class_size)
head(stage1.long)
##       ID_unit class class_size stage1prob studentid
## 1           1     1         44  0.3400309         1
## 1.1         1     1         44  0.3400309         2
## 1.2         1     1         44  0.3400309         3
## 1.3         1     1         44  0.3400309         4
## 1.4         1     1         44  0.3400309         5
## 1.5         1     1         44  0.3400309         6
# use strata function to select 4 ssus from each psu
stage2.units<-strata(stage1.long, stratanames=c("class"),
                size=rep(4,5), method="srswor")
nrow(stage2.units)
## [1] 20
# get the data for the second-stage sample
ssusample<-getdata(stage1.long,stage2.units)
head(ssusample)
##       class_size stage1prob studentid class ID_unit       Prob Stratum
## 1.3          44  0.3400309         4     1       4 0.09090909       1
## 1.13         44  0.3400309        14     1      14 0.09090909       1
## 1.21         44  0.3400309        22     1      22 0.09090909       1
## 1.26         44  0.3400309        27     1      27 0.09090909       1
## 3.11         26  0.2009274        12     3      56 0.15384615       2
## 3.18         26  0.2009274        19     3      63 0.15384615       2
```

The last step is computing the final selection probability, accounting for both stages of sampling, and the final sampling weight. In data frame *ssusample*, variable *stage1prob* contains the psu-level sampling probability (we defined this variable earlier and used it to select the psus) and variable *Prob* contains the ssu-level sampling probability (this is computed by the *strata* function and for this example equals $4/M_i$). Thus, the selection probability for each student in the sample is the product *stage1prob* × *Prob*. The final weight is the reciprocal of the final selection probability, which for this example equals 32.35 for all students in the sample because the first-stage sample was selected with probability proportional to $M_i$ and the same subsample size (here, $m_i = 4$) was selected from each psu.

```
# compute the sampling weights
# stage1prob contains stage 1 sampling probability;
# Prob has stage 2 sampling probability
ssusample$finalprob<- ssusample$stage1prob*ssusample$Prob
ssusample$finalwt<-1/ssusample$finalprob
sum(ssusample$finalwt)  # check sum of weights
## [1] 647
# print selected columns of ssusample
print(ssusample[,c(1,2,3,4,6,8,9)],digits=4)
##       class_size stage1prob studentid class    Prob finalprob finalwt
## 1.3          44     0.3400         4     1 0.09091   0.03091   32.35
## 1.13         44     0.3400        14     1 0.09091   0.03091   32.35
## 1.21         44     0.3400        22     1 0.09091   0.03091   32.35
## 1.26         44     0.3400        27     1 0.09091   0.03091   32.35
## 3.11         26     0.2009        12     3 0.15385   0.03091   32.35
## 3.18         26     0.2009        19     3 0.15385   0.03091   32.35
## 3.19         26     0.2009        20     3 0.15385   0.03091   32.35
## 3.24         26     0.2009        25     3 0.15385   0.03091   32.35
## 7.11         20     0.1546        12     7 0.20000   0.03091   32.35
## 7.13         20     0.1546        14     7 0.20000   0.03091   32.35
## 7.18         20     0.1546        19     7 0.20000   0.03091   32.35
## 7.19         20     0.1546        20     7 0.20000   0.03091   32.35
## 13.16        46     0.3555        17    13 0.08696   0.03091   32.35
## 13.31        46     0.3555        32    13 0.08696   0.03091   32.35
## 13.34        46     0.3555        35    13 0.08696   0.03091   32.35
## 13.42        46     0.3555        43    13 0.08696   0.03091   32.35
## 14.1        100     0.7728         2    14 0.04000   0.03091   32.35
## 14.20       100     0.7728        21    14 0.04000   0.03091   32.35
## 14.35       100     0.7728        36    14 0.04000   0.03091   32.35
## 14.68       100     0.7728        69    14 0.04000   0.03091   32.35
```

## 6.3   Computing Estimates from an Unequal-Probability Sample

The syntax used to compute estimates from an unequal-probability cluster sample is largely the same as that used in Chapter 5 for equal-probability cluster samples. The *svymean* and *svytotal* functions of the **survey** package (Lumley, 2020) calculate estimates of means, totals, and proportions by using the formulas with survey weights. When the *fpc* argument is omitted from the *svydesign* function call, standard errors are calculated with the formulas for the with-replacement variance in Section 6.4 of SDA.

### 6.3.1   Estimates from with-Replacement Samples

**Example 6.4 of SDA.** This example shows how to calculate estimates when the cluster total $t_i$ has already been found for each psu (or when the psus are also the observation units, that is, $M_i = 1$ for all psus). Since the summary statistic has already been calculated for each psu, the *svydesign* function is called with `id=~1`. We only need to specify the unequal weights using the *weights* argument to calculate the estimates. Class 14 appears twice in the data since it was selected twice for the sample—we call it class 141 for the first appearance and class 142 for the second to distinguish them.

The mean calculated from *svymean* estimates $\bar{t}_{\mathcal{U}}$, the population mean of the cluster totals $t_i$, which for this example is the total amount of time spent studying by students in class $i$. The average amount of time spent studying per student is estimated by the ratio $\hat{\bar{y}}_\psi = \hat{\bar{t}}_\psi / \hat{\bar{M}}_\psi$. The *svyratio* function can give the estimate $\hat{\bar{y}}_\psi$. (If the data set consists of the individual values $y_{ij}$ instead of the summary statistics, then the mean $\hat{\bar{y}}_\psi$ can be estimated directly from *svymean*, as seen in the code for Example 6.6 of SDA.)

```
studystat <- data.frame(class = c(12, 141, 142, 5, 1),
                        Mi = c(24, 100, 100, 76, 44),
                        tothours=c(75,203,203,191,168))
studystat$wt<-647/(studystat$Mi*5)
sum(studystat$wt) # check weight sum, which estimates N=15 psus
## [1] 12.62321
# design for with-replacement sample, no fpc argument
d0604 <- svydesign(id = ~1, weights=~wt, data = studystat)
d0604
## Independent Sampling design (with replacement)
## svydesign(id = ~1, weights = ~wt, data = studystat)
# Ratio estimation using Mi as auxiliary variable
ratio0604<-svyratio(~tothours, ~Mi,design = d0604)
ratio0604
## Ratio estimator: svyratio.survey.design2(~tothours, ~Mi, design = d0604)
## Ratios=
##                 Mi
## tothours 2.703268
## SEs=
##                 Mi
## tothours 0.3437741
confint(ratio0604, level=.95,df=4)
##                  2.5 %   97.5 %
## tothours/Mi 1.748798 3.657738
# Can also estimate total hours studied for all students in population
svytotal(~tothours,d0604)
##            total    SE
## tothours   1749 222.42
```

The average amount of time a student spent studying statistics is estimated as 2.70 hours with an estimated standard error of 0.34 hours and a 95% confidence interval of [1.74, 3.66]. Note that 4 degrees of freedom (df; here, 1 less than the number of psus) are used for the confidence interval.

**Example 6.6 of SDA.** The estimates for a two-stage cluster sample with replacement are calculated exactly the same way as for a one-stage sample. For this example, we have data for the individual students in the psus so we enter those for each student.

Class 14 appears twice in the sample of psus in Example 6.4 of SDA. An independent set of students is selected for each appearance. To enable correct variance calculations, the first occurrence of class 14 is relabeled as class 141, and the second occurrence as class 142. These are counted as two separate psus in the estimation. If you labeled both as 14, then the *id* argument of *svydesign* would treat that as one psu with $m_i = 10$ instead of two psus of size 5.

The weight *studentwt* is calculated as the first-stage weight $M_0/(nM_i)$ times the second-stage weight $M_i/m_i$. The sample is self-weighting and the weight for each student simplifies to $647/25$. For many problems, defining the weights is the trickiest part, and it is also the most important. Always check that the sum of the weights approximately (or exactly, in this case) equals the population size.

```
students <- data.frame(class = rep(studystat$class,each=5),
   popMi = rep(studystat$Mi,each=5),
   sampmi=rep(5,25),
   hours=c(2,3,2.5,3,1.5,2.5,2,3,0,0.5,3,0.5,1.5,2,3,1,2.5,3,5,2.5,4,4.5,3,2,5))
# The 'with' function allows us to calculate using variables from a data frame
# without having to type the data frame name for all of them
students$studentwt <- with(students,(647/(popMi*5)) * (popMi/sampmi))
# check the sum of the weights
sum(students$studentwt)
## [1] 647
# create the design object
d0606 <- svydesign(id = ~class, weights=~studentwt, data = students)
d0606
## 1 - level Cluster Sampling design (with replacement)
## With (5) clusters.
## svydesign(id = ~class, weights = ~studentwt, data = students)
# estimate mean and SE
svymean(~hours,d0606)
##        mean      SE
## hours   2.5 0.3606
degf(d0606)
## [1] 4
confint(svymean(~hours,d0606),level=.95,df=4) #use t-approximation
##          2.5 %   97.5 %
## hours 1.498938 3.501062
# estimate total and SE
svytotal(~hours,d0606)
##         total      SE
## hours  1617.5 233.28
confint(svytotal(~hours,d0606),level=.95,df=4)
##          2.5 %   97.5 %
## hours 969.8132 2265.187
```

In the *svydesign* function, we supply the weights (which, for this example, are the same for all students) but no *fpc* argument. We specify the psu membership in the *id* argument. This means that the variability is calculated at the first stage level using the pps with-replacement formulas, that is, the variability among $\hat{t}_i/\psi_i$. Note that 4 df (1 less than the number of psus) are used for the confidence interval.

## 6.3.2 Estimates from without-Replacement Samples

Even when an unequal-probability sample was selected without replacement, the with-replacement variance is commonly calculated for simplicity and stability. Use the *weights* argument to provide the sampling weights at the observation-unit level, and use the `id=~psuid` argument to provide the information on psu membership (recall that only the psu membership is needed to calculate the with-replacement variance).

**Example 6.11 of SDA.** This example analyzes the without-replacement unequal-probability sample the same way as for the sample in Example 6.6. Even though the sample was selected without replacement, the with-replacement variance provides a good approximation.

```
data(classpps)
nrow(classpps)
## [1] 20
head(classpps)
##   class class_size finalweight hours
## 1     4         22       32.35   5.0
## 2     4         22       32.35   4.5
## 3     4         22       32.35   5.5
## 4     4         22       32.35   5.0
## 5    10         34       32.35   2.0
## 6    10         34       32.35   4.0
d0611 <- svydesign(ids = ~class, weights=~classpps$finalweight, data = classpps)
d0611
## 1 - level Cluster Sampling design (with replacement)
## With (5) clusters.
## svydesign(ids = ~class, weights = ~classpps$finalweight, data = classpps)
# estimate mean and SE
svymean(~hours,d0611)
##        mean     SE
## hours 3.45 0.4819
confint(svymean(~hours,d0611),level=.95,df=4) #use t-approximation
##           2.5 %   97.5 %
## hours 2.112147 4.787853
# estimate total and SE
svytotal(~hours,d0611)
##        total     SE
## hours 2232.2 311.76
confint(svytotal(~hours,d0611),level=.95,df=4)
##          2.5 %   97.5 %
## hours 1366.559 3097.741
```

**Calculating the without-replacement variance for a one-stage sample.** In general, we recommend calculating the with-replacement variance estimate and omitting the *fpc* argument from *svydesign* when unequal-probability sampling is used. Most of the replication methods for calculating variances in Chapter 9 also calculate with-replacement variances. You can skip the remainder of this section if the with-replacement variances work for your applications.

Functions in the **survey** package will calculate the without-replacement variance for some one-stage designs if you specify the inclusion probabilities in the *fpc* argument. (As of this writing, the package will not yet calculate without-replacement variances for two-stage designs—the situation where unequal-probability sampling is most commonly used.) The formulas for calculating the Horvitz-Thompson (HT), Sen-Yates-Grundy (SYG), and other without-replacement variance estimates require knowledge of the joint inclusion probabilities, so you must also supply those to the *svydesign* function.

Let's calculate some joint inclusion probabilities first and then use them in the *svydesign* function. Functions in the **sampling** package will calculate joint inclusion probabilities for some of the sample-selection methods; for example, function *UPsampfordpi2* will calculate the joint inclusion probabilities for a sample selected using Sampford's method. For some other sample-selection methods, however, the joint inclusion probabilities must be calculated directly.

**Example 6.8 of SDA.** We use the supermarket example to illustrate the calculation of joint inclusion probabilities when the sample size is 2. We first create a data frame of the supermarket population containing the store identifier, area, and revenue.

```
supermarket<-data.frame(store=c('A','B','C','D'),area=c(100,200,300,1000),
                        ti=c(11,20,24,245))
supermarket
##   store area  ti
## 1     A  100  11
## 2     B  200  20
## 3     C  300  24
## 4     D 1000 245
```

The draw-by-draw method was used to select two supermarkets for the sample, where the selection probability for draw 1 was proportional to store area. We can use that information to calculate $\pi_i$, the probability of store $i$ being included in the sample, and $\pi_{ik}$, the joint probability that stores $i$ and $k$ are both included in the sample.

Here, we use matrix operations to calculate the probabilities by applying the formulas in Example 6.8 of SDA, noting that if **a** and **b** are two vectors, the $(i,j)$ entry of $\mathbf{ab}^T$ is $a_i b_j$. The *apply* function sums the entries in each column.

```
supermarket$psi<-supermarket$area/sum(supermarket$area)
psii<-supermarket$area/sum(supermarket$area)
piik<- psii %*% t(psii/(1-psii)) + (psii/(1-psii)) %*% t(psii)
diag(piik)<-rep(0,4) # set the diagonal entries of the matrix equal to zero
piik  # joint inclusion probabilities
##            [,1]       [,2]       [,3]      [,4]
## [1,] 0.00000000 0.01726190 0.02692308 0.1458333
## [2,] 0.01726190 0.00000000 0.05563187 0.2976190
## [3,] 0.02692308 0.05563187 0.00000000 0.4567308
## [4,] 0.14583333 0.29761905 0.45673077 0.0000000
pii<-apply(piik,2,sum)
pii # inclusion probabilities
## [1] 0.1900183 0.3705128 0.5392857 0.9001832
```

The results show that $\pi_1 = 0.19$, $\pi_2 = 0.37$, $\pi_3 = 0.539$, and $\pi_4 = 0.90$. The joint inclusion probabilities are given in *piik*: for example, $\pi_{12} = 0.01726$. These are the numbers shown in Table 6.6 of SDA.

Now let's use the values of $\pi_i$ and $\pi_{ik}$ to calculate the HT and SYG variance estimates. Of course, since the supermarket sample has only two units, neither estimate will be very accurate, but it will serve to illustrate the methods.

**Example 6.9 of SDA.** Suppose supermarkets C and D were selected from the population in Example 6.8 of SDA. We will calculate the Horvitz–Thompson (HT) estimate for total revenue and the without-replacement HT and SYG variance estimates.

As always, we specify all the information about the design in the *svydesign* function. We tell the function that this is an unequal-probability sample without replacement through the *fpc* argument. Instead of giving the population sizes in the *fpc* argument, however, for pps sampling without replacement we specify `fpc=~pii`, the inclusion probability for each sampled unit.

We also use two other arguments to the *svydesign* function that we have not seen before. The *variance* argument tells whether to calculate the HT or SYG (the function calls this "YG") formula for the variance. We supply the joint inclusion probabilities in the *pps* argument, after first placing $\pi_i$ on the diagonal elements of the joint probabilities matrix and using the function *ppsmat* to get the joint probabilities in the form required by *svydesign*.

```
supermarket2<-supermarket[3:4,]
supermarket2$pii <- pii[3:4] # these are the unit inclusion probs when n=2
jointprob<-piik[3:4,3:4]  # joint probability matrix for stores C and D
diag(jointprob)<-supermarket2$pii # set diagonal entries equal to pii
jointprob
##           [,1]      [,2]
## [1,] 0.5392857 0.4567308
## [2,] 0.4567308 0.9001832
# Horvitz-Thompson type
dht<- svydesign(id=~1,  fpc=~pii, data=supermarket2,
                pps=ppsmat(jointprob),variance="HT")
dht
## Sparse-matrix design object:
##  svydesign(id = ~1, fpc = ~pii, data = supermarket2, pps = ppsmat(jointprob),
##      variance = "HT")
svytotal(~ti,dht)
##     total      SE
## ti 316.67 82.358
# Sen-Yates-Grundy type
dsyg<- svydesign(id=~1,  fpc=~pii, data=supermarket2,
                 pps=ppsmat(jointprob),variance="YG")
dsyg
## Sparse-matrix design object:
##  svydesign(id = ~1, fpc = ~pii, data = supermarket2, pps = ppsmat(jointprob),
##      variance = "YG")
svytotal(~ti,dsyg)
##     total      SE
## ti 316.67 57.094
```

We can compare the variance estimates from the two methods with the true without-replacement variance $V(\hat{t}_{\text{HT}}) = 4383.6$ in SDA (which is known for this small example where the full population is known), with $\hat{V}_{\text{HT}}(\hat{t}_{\text{HT}}) = (82.358)^2 = 6782.8$, and $\hat{V}_{\text{SYG}}(\hat{t}_{\text{HT}}) = (57.094)^2 = 3259.8$. In most situations, the SYG variance estimate is preferred because it is more stable.

The *svydesign* function also provides some approximation methods to calculate without-replacement variance estimates for one-stage samples. Option `pps=HR(sum(piisq)/n)`, where *piisq* is the vector of squared inclusion probabilities and $n$ is the number of psus selected, gives the Hartley and Rao (1962; see Exercise 6.36 in SDA) approximation to the variance.

**Example 6.10 of SDA.** Let's do one more example, to compare the with-replacement, HT, and SYG variance estimates calculated for the unequal-probability sample in data *agpps*, as well as the Hartley–Rao approximation. We would expect the with-replacement variance estimate to work well here because $n = 15$ is small relative to $N = 3078$.

```
data(agpps)
jtprobag<-as.matrix(agpps[,20:34])
diag(jtprobag)<-agpps$SelectionProb
# Horvitz-Thompson type
dhtag<- svydesign(id=~1,  fpc=~SelectionProb, data=agpps,
                  pps=ppsmat(jtprobag),variance="HT")
svytotal(~acres92,dhtag)
```

```
##               total         SE
## acres92 936291172 70466858
# Sen-Yates-Grundy type
dsygag<- svydesign(id=~1,  fpc=~SelectionProb, data=agpps,
                   pps=ppsmat(jtprobag),variance="YG")
svytotal(~acres92,dsygag)
##               total         SE
## acres92 936291172 11715201
# Hartley-Rao approximation
sumsqprob<-sum(agpps$SelectionProb^2)/nrow(agpps)
dHRag<-svydesign(id=~1, fpc=~SelectionProb, data=agpps,
                 pps=HR(sumsqprob),variance="YG")
svytotal(~acres92,dHRag)
##               total         SE
## acres92 936291172 12148234
# With-replacement variance
dwrag<-svydesign(id=~1, weights=~SamplingWeight, data=agpps)
svytotal(~acres92,dwrag)
##               total         SE
## acres92 936291172 12293009
```

Note that the with-replacement (12,293,009), SYG (11,715,201), and Hartley–Rao (12,148,234) standard errors are all similar to each other. The HT standard error (70,466,858) is larger and often less stable; in general, we recommend one of the other methods.

## 6.4  Summary, Tips, and Warnings

Several functions in the `sampling` package will select equal- and unequal-probability cluster samples; some of these are listed in Table 6.1. The *sample* function can be used to select with-replacement unequal-probability samples.

Table 6.2 lists the major R functions used in this chapter.

**Tips and Warnings**

- When selecting an unequal-probability sample, check the calculation of the selection probabilities to make sure these are roughly proportional to the unit sizes.

- The more complex the sampling plan, the more complicated the weight calculations. Check that the sum of the weights approximately equals the population size.

- For unequal-probability sampling, omitting the *fpc* argument in the *svydesign* function gives the with-replacement variance. In general, this is the approach that we recommend. If the without-replacement variance is desired, use the Sen-Yates-Grundy formula directly.

**TABLE 6.2**

Functions used for Chapter 6.

| Function | Package | Usage |
|---|---|---|
| sample | base | Select a with-replacement sample with unequal probabilities |
| confint | stats | Calculate confidence intervals, add df for $t$ confidence interval |
| apply | base | Apply a function to the rows or columns of a matrix |
| cluster | sampling | Select a cluster sample |
| strata | sampling | Select a stratified random sample (here used to select ssus from the sampled psus) |
| mstage | sampling | Select a multi-stage cluster sample |
| UPsampford | sampling | Select an unequal-probability sample of units using Sampford's method |
| inclusionprobabilities | sampling | Convert a vector of positive size measures to selection probabilities, for use in the UP selection functions |
| getdata | sampling | Extract the data after selecting a sample |
| svydesign | survey | Specify the survey design |
| svymean | survey | Calculate mean and standard error of mean |
| svyratio | survey | Calculate ratio and standard error of ratio |
| svytotal | survey | Calculate total and standard error of total |

# 7

## *Complex Surveys*

We have already seen most of the components needed for selecting and computing estimates from a stratified multistage sample. Now let's put them all together. Section 7.1 shows how to select a stratified two-stage sample, and Section 7.3 shows how to compute estimates using examples from the National Health and Nutrition Examination Survey (NHANES).

The new features considered in this chapter are how to estimate quantiles (Section 7.2) and how to graph survey data (Sections 7.4 and 7.5). The code is in file `ch07.R` on the book website.

## 7.1 Selecting a Stratified Two-Stage Sample

In the following example, we select a sample of classes from the small population considered in Section 6.2 of SDA, after first dividing the classes into three strata based on their sizes. Stratum 1 contains the two large classes, stratum 2 contains six medium-sized classes, and stratum 3 contains the seven smallest classes. The code specifies drawing two primary sampling units (psus) without replacement (srswor) from each stratum, and drawing three students without replacement from each selected class. Of course other allocations of psus to strata can be used, as described in Chapter 3. The psus are arranged in strata, so the only new feature here is to add the stratification information in the function *mstage* from the **sampling** package (Tillé and Matei, 2021).

We use the *classeslong* data frame that we created in Section 1.5, where variable *class* gives the psu number and variable *studentid* gives the student identifier (numbered from 1 to *class_size*) within each class.

```
# data(classeslong)
# define strata
classeslong$strat<-rep(3,nrow(classeslong))
classeslong$strat[classeslong$class_size > 40]<-2
classeslong$strat[classeslong$class_size > 70]<-1
# table(classeslong$class,classeslong$strat)
# order data by stratum
classeslong2<-classeslong[order(classeslong$strat),]
# check the stratum construction
table(classeslong2$strat,classeslong2$class_size)
##
##      15  20  22  24  26  33  34  44  46  54  63  76 100
##  1    0   0   0   0   0   0   0   0   0   0   0  76 100
##  2    0   0   0   0   0   0   0  88  92  54  63   0   0
##  3   15  20  22  24  26  33  34   0   0   0   0   0   0
nrow(classeslong2) # number of students in population
## [1] 647
```

DOI: 10.1201/9781003228196-7

To select a stratified cluster sample using *mstage*, we first define *numberselect* consisting of information on stratum sizes, number of psus selected from each stratum, and number of secondary sampling units (ssus; here, students) selected within each sampled psu as follows:

```
numberselect<-list(table(classeslong2$strat),rep(2,3),rep(3,6))
```

Next, use the *mstage* function to select a stratified two-stage cluster sample.

```
mstage(classeslong2,stage=list("stratified","cluster","stratified"),
            varnames=list("strat","class","studentid"),
            size=numberselect, method=list("","srswor","srswor"))
```

The *stage* argument for this example is a list with three components. The first component, `"stratified"`, defines the stratification for the psus, but nothing is selected at this stage (the first component of *method* is blank). Then, an SRS of psus (variable *class*) is selected within each stratum (this is described by the second components of *stage*, *varnames*, and *method*). At this point, the sample consists of the selected psus from the 3 strata. We then use `"stratified"` again to select an SRS of students from each sampled psu.

The result from *mstage* is a list with three components corresponding to the components of *stage* in the function argument. The third component is the final sample, saved in *sample3*. The variable *Prob*, computed by function *mstage*, is the final selection probability, and we calculate *finalweight* as its reciprocal.

```
# select a stratified two stage cluster sample,
# stratum: strat
# psu: class
# ssu: studentid
# number of psus selected n =2, size=rep(n=2,3 strata) (srswor)
# number of students selected m_i =3 size=rep(m_i= 3,6 classes) (srswor)
numberselect<-list(table(classeslong2$strat),rep(2,3),rep(3,6))
numberselect
## [[1]]
##
##   1   2   3
## 176 297 174
##
## [[2]]
## [1] 2 2 2
##
## [[3]]
## [1] 3 3 3 3 3 3
# select a stratified two-stage cluster sample
set.seed(75745)
tempid<-mstage(classeslong2,stage=list("stratified","cluster","stratified"),
            varnames=list("strat","class","studentid"),
            size=numberselect, method=list("","srswor","srswor"))
# get data
sample3<-getdata(classeslong2,tempid)[[3]]   #3rd stage
sample3$finalweight<-1/sample3$Prob
# check sum of weights, should be close to number of students in population
# (but not exactly equal, since psus not selected with prob proportional to M_i)
sum(sample3$finalweight)
## [1] 624
sample3 # print the sample
##        class class_size studentid strat ID_unit Prob_ 3 _stage        Prob
## 5.31       5         76        32     1      32    0.03947368 0.03947368
```

```
## 5.42      5         76          43      1      43     0.03947368 0.03947368
## 5.61      5         76          62      1      62     0.03947368 0.03947368
## 14.9     14        100          10      1      86     0.03000000 0.03000000
## 14.37    14        100          38      1     114     0.03000000 0.03000000
## 14.79    14        100          80      1     156     0.03000000 0.03000000
## 8.34      8         44          35      2     318     0.06818182 0.02272727
## 8.39      8         44          40      2     323     0.06818182 0.02272727
## 8.28      8         44          29      2     312     0.06818182 0.02272727
## 9.26      9         54          27      2     354     0.05555556 0.01851852
## 9.28      9         54          29      2     356     0.05555556 0.01851852
## 9.38      9         54          39      2     366     0.05555556 0.01851852
## 7.17      7         20          18      3     572     0.15000000 0.04285714
## 7.5       7         20           6      3     560     0.15000000 0.04285714
## 7.10      7         20          11      3     565     0.15000000 0.04285714
## 12.6     12         24           7      3     615     0.12500000 0.03571429
## 12.9     12         24          10      3     618     0.12500000 0.03571429
## 12.16    12         24          17      3     625     0.12500000 0.03571429
##            finalweight
## 5.31       25.33333
## 5.42       25.33333
## 5.61       25.33333
## 14.9       33.33333
## 14.37      33.33333
## 14.79      33.33333
## 8.34       44.00000
## 8.39       44.00000
## 8.28       44.00000
## 9.26       54.00000
## 9.28       54.00000
## 9.38       54.00000
## 7.17       23.33333
## 7.5        23.33333
## 7.10       23.33333
## 12.6       28.00000
## 12.9       28.00000
## 12.16      28.00000
```

You can also look at the selection probabilities for each stage if desired. The following extracts the other stages in *sample1* and *sample2*. The first stage defines the stratification, so Prob_ 1 _stage is 1. Stratum 1 contains two psus, so the function selects each with certainty. Classes are selected from stratum 2 with probability of 2/6, and classes are selected from stratum 3 with probability of 2/7. These are reported as Prob_ 2 _stage of *sample2*. At the third stage, Prob_ 3 _stage is calculated as $3/class\_size$. The final probability *Prob* is calculated as 1*Prob_ 2 _stage*Prob_ 3 _stage.

```
sample1<-getdata(classeslong2,tempid)[[1]]  #1st stage
sample2<-getdata(classeslong2,tempid)[[2]]  #2nd stage
names(sample1)
## [1] "class"           "class_size"     "studentid"        "strat"
## [5] "ID_unit"         "Prob_ 1 _stage" "Stratum"
table(sample1$`Prob_ 1 _stage`)
##
##   1
## 647
table(sample2$strat,sample2$`Prob_ 2 _stage`) # Selection probs for psus in strata
```

```
##
##      0.285714285714286 0.333333333333333   1
## 1                    0                   0 176
## 2                    0                  98 0
## 3                   44                   0 0
table(sample3$class,sample3$`Prob_ 3 _stage`) # Selection probs for ssus in psus
##
##      0.03 0.0394736842105263 0.0555555555555556 0.0681818181818182 0.125 0.15
## 5    0                     3                  0                  0     0    0
## 7    0                     0                  0                  0     0    3
## 8    0                     0                  0                  3     0    0
## 9    0                     0                  3                  0     0    0
## 12   0                     0                  0                  0     3    0
## 14   3                     0                  0                  0     0    0
```

For more complicated designs, you may want to select the sample at each stage separately, as illustrated in Section 6.2. For example, you can use function *UPsampford* to select a sample of psus from each stratum, then select the sample at the subsequent stages.

## 7.2   Estimating Quantiles

Quantiles are estimated using the empirical cumulative distribution function (cdf) $\hat{F}(y)$, which is the sum of the weights for the sample observations having $y_i \leq y$ divided by the sum of all of the weights. Because $\hat{F}(y)$ has jumps at the distinct values of $y$ in the sample, however, for many values of $q$ there is no value of $y$ in the sample that has $\hat{F}(y)$ exactly equal to $q$. Multiple definitions for population and sample quantiles have been proposed (Hyndman and Fan, 1996; Wang, 2021).

The *svyquantile* function in the **survey** package (Lumley, 2020) calculates several estimates of quantiles and their standard errors. With `ties="discrete"`, the empirical cdf $\hat{F}$ is used directly, with jumps at the values of $y$ in the sample. With `ties="rounded"` an interpolated cdf is used (see Exercise 7.19 of SDA). We usually prefer interpolated quantiles, as they smooth out an empirical cdf that has large jumps.

The quantiles $\tilde{\theta}_q$ are calculated by requesting the desired quantile values in the *svyquantile* function. Request the 0.25, 0.5, 0.75, and 0.90 quantiles, for example, by typing `quantiles=c(0.25, 0.50, 0.75, 0.90)`.

**Example 7.5 of SDA.** The following code requests quantiles and CIs for the height values in the SRS *htsrs*. Of course, for a complex design, one would include stratification and clustering information in the *svydesign* function.

```
data(htsrs)
dhtsrs<-svydesign(id = ~1,weights=rep(2000/200,200),fpc=rep(2000,200), data=htsrs)
# cdf treated as step function, gives values in Table 7.1 of SDA
svyquantile(~height, dhtsrs, quantiles=c(0.25,0.5,0.75,0.9), ties = "discrete")
##        0.25 0.5 0.75 0.9
## height 160 169  176 184
# interpolated quantiles (usually preferred method)
svyquantile(~height, dhtsrs, quantiles=c(0.25,0.5,0.75,0.9), ties = "rounded")
##         0.25    0.5 0.75   0.9
## height 159.7 168.75  176 183.4
```

The *svyquantile* function will also calculate confidence intervals for quantiles if you request ci=TRUE. The default method is `interval.type="Wald"`, which calculates the Woodruff (1952) interval presented in Section 9.5 of SDA.

**Examples 7.6 and 9.12 of SDA.** Here we calculate confidence intervals for the interpolated quantiles in the *htstrat* data.

```
data(htstrat)
popsize_recode <- c('F' = 1000, 'M' = 1000)
# create a new variable popsize for population size
htstrat$popsize<-popsize_recode[htstrat$gender]
head(as.data.frame(htstrat))
##      rn height gender popsize
## 1 201    166      F    1000
## 2 965    163      F    1000
## 3 490    166      F    1000
## 4 249    155      F    1000
## 5 260    154      F    1000
## 6 324    160      F    1000
# design object
# svydesign calculates the weights here from the fpc argument
dhtstrat<-svydesign(id = ~1, strata = ~gender, fpc = ~popsize,
          data = htstrat)
# ties = "discrete" gives values in Table 7.1 of SDA
svyquantile(~height, dhtstrat, c(0.25,0.5,0.75,0.9), ties = "discrete")
##         0.25 0.5 0.75 0.9
## height   161 168  177 182
# ties = "rounded" gives values in Example 9.12 of SDA
svyquantile(~height, dhtstrat, c(0.25,0.5,0.75,0.9), ties = "rounded",
            ci=TRUE, interval.type = "Wald")
## $quantiles
##             0.25      0.5    0.75   0.9
## height 160.7143 167.5556 176.625 181.5
##
## $CIs
## , , height
##
##             0.25      0.5     0.75      0.9
## (lower 159.3556 165.8078 173.3572 178.7176
## upper) 162.0247 170.0942 178.5439 190.1679
```

## 7.3 Computing Estimates from Stratified Multistage Samples

We have seen all the building blocks for computing the estimates from any survey. Now let's put them all together using the data from the National Health and Nutrition Examination Survey (NHANES, Centers for Disease Control and Prevention, 2017). The Centers for Disease Control and Prevention produce online tutorials for analyzing NHANES data. These, and sample code and tips for analyzing NHANES data using R, can be found at https://wwwn.cdc.gov/nchs/nhanes/tutorials/.

**Example 7.9 of SDA.** In this example, we look at statistics about body mass index (BMI, variable *bmxbmi*) for adults age 20 and over. We will compute these estimates using *svymean* and *svyquantile*.

One of the statistics to be calculated is the proportion of adults having BMI greater than 30, so we define categorical variable *bmi30* to equal 1 if the person's BMI is greater than 30 and 0 if it is less than or equal to 30. Note that the variable *bmxbmi* has missing values, so we set *bmi30* to be missing if *bmxbmi* is missing. In the data file `nhanes.csv`, missing values are coded by −9. We coded all the missing values −9 as "NA" in the R data set *nhanes* in the **SDAResources** package.

We also need to define a variable giving the domain of interest to be analyzed: adults age 20 and over who have data for *bmxbmi*. We define *age20d*=1 if *ridageyr*≥ 20 and *bmxbmi* is not missing, and 0 otherwise. This excludes the observations with missing values from the domain of interest, and ensures that standard errors for the domain are calculated correctly.

```
data(nhanes)
nrow(nhanes) #9971
## [1] 9971
names(nhanes)
##  [1] "sdmvstra" "sdmvpsu" "wtint2yr" "wtmec2yr" "ridstatr" "ridageyr"
##  [7] "ridagemn" "riagendr" "ridreth3" "dmdeduc2" "dmdfmsiz" "indfmpir"
## [13] "bmxwt"    "bmxht"    "bmxbmi"   "bmxwaist" "bmxleg"   "bmxarml"
## [19] "bmxarmc"  "bmdavsad" "lbxtc"    "bpxpls"   "sbp"      "dbp"
## [25] "bpread"
# count number of observations with missing value for ridageyr, bmxbmi
sum(is.na(nhanes$ridageyr)) # ridageyr gives age in years
## [1] 0
sum(is.na(nhanes$bmxbmi))   # bmxbmi gives BMI
## [1] 1215
# define age20d and bmi30
nhanes$age20d<-rep(0,nrow(nhanes))
nhanes$age20d[nhanes$ridageyr >=20 & !is.na(nhanes$bmxbmi)]<-1
nhanes$bmi30<-nhanes$bmxbmi
nhanes$bmi30[nhanes$bmxbmi>30]<-1
nhanes$bmi30[nhanes$bmxbmi<=30]<-0
nhanes$bmi30<-factor(nhanes$bmi30) # set bmi30 as a categorical variable
# check missing value counts for new variables
sum(is.na(nhanes$age20d))
## [1] 0
sum(is.na(nhanes$bmi30))
## [1] 1215
sum(nhanes$age20d) # how many records in domain?
## [1] 5406
head(nhanes)
##   sdmvstra sdmvpsu  wtint2yr  wtmec2yr ridstatr ridageyr ridagemn riagendr
## 1      125       1 134671.37 135629.51        2       62       NA        1
## 2      125       1  24328.56  25282.43        2       53       NA        1
## 3      131       1  12400.01  12575.84        2       78       NA        1
## 4      131       1 102718.00 102078.63        2       56       NA        2
## 5      126       2  17627.67  18234.74        2       42       NA        2
## 6      128       1  11252.31  10878.68        2       72       NA        2
##   ridreth3 dmdeduc2 dmdfmsiz indfmpir bmxwt bmxht bmxbmi bmxwaist bmxleg
## 1        3        5        2     4.39  94.8 184.5   27.8    101.1   43.3
## 2        3        3        1     1.32  90.4 171.4   30.8    107.9   38.0
```

```
## 3          3         3        2    1.51  83.4 170.1  28.8     116.5   35.6
## 4          3         5        1    5.00 109.8 160.9  42.4     110.1   38.5
## 5          4         4        5    1.23  55.2 164.9  20.3      80.4   37.4
## 6          1         2        5    2.82  64.4 150.0  28.6      92.9   34.4
##    bmxarml bmxarmc bmdavsad lbxtc bpxpls sbp dbp bpread age20d bmi30
## 1     43.6    35.9     22.8   173     76 120  63      2      1     0
## 2,    40.0    33.2     27.3   265     72 137  85      2      1     1
## 3     37.0    31.0     26.6   229     56 134  45      2      1     0
## 4     37.7    38.3     25.1   174     78 135  69      2      1     1·
## 5     36.0    27.2       NA   204     76 106  55      2      1     0
## 6     33.5    31.4     23.1   190     64 121  59      2      1     0
```

Next, we use *svydesign* to describe the design information and use *subset* to select the domain of adults with data for BMI for analysis.

```
# stratified cluster design
d0709 <- svydesign(id = ~sdmvpsu, strata=~sdmvstra, weights=~wtmec2yr,
                 nest=TRUE, data = nhanes)
# domain estimation, age20+
d0709sub<-subset(d0709, age20d ==1)
d0709sub
## Stratified 1 - level Cluster Sampling design (with replacement)
## With (30) clusters.
## subset(d0709, age20d == 1)
```

Let's look at the features used in the *svydesign* function to describe the design.

- The *weights* argument in *svydesign* specifies the variable containing the final weights. The data set contains two weight variables: *wtint2yr* gives the weight for the set of persons with interview data, and *wtmec2yr* gives the weight for the subset of interviewed persons who had a medical examination. BMI is measured in the medical examination, so the appropriate weight variable to use is *wtmec2yr*.

- The *strata* and *id* arguments are used exactly as in Chapters 3 and 5, except now we include both of them. The `strata=~sdmvstra` argument says that *sdmvstra* is the variable giving the stratum membership. The psus specified in `id=~sdmvpsu` are the first-stage sampling units.

- In the NHANES data, the two psus in each stratum are labeled as '1' and '2'. The `nest=TRUE` argument says that psu labels are nested within strata—that is, multiple strata have the same psu labels. Typing `nest=TRUE` ensures that psu 1 in stratum 1 is recognized as being a different psu than psu 1 in stratum 2.

  When there is one stratum, the results are the same if you have `nest=TRUE`, `nest=FALSE`, or simply omit the *nest* argument.

- No *fpc* argument is included in *svydesign*. With complex samples such as NHANES, we usually want to calculate the with-replacement variance, which requires only psu-level information.

The *subset* function to define domains has been discussed in Chapter 4. This specifies that estimates are desired for the domain of persons age 20 and older having data for BMI (with *age20d*=1), and carries the stratification and clustering information from the full design over for analyzing the subset. If you just created a subset of the data consisting of the observations having *ridageyr* $\geq$ 20, in some instances (for example, when some psus have no members of the domain), the standard errors would be incorrect; by using the *subset* function, the correct standard errors are calculated.

Note that we exclude adults with missing values of *bmxbmi* from the domain of interest with *age20d*=1. The estimates are computed from the adults who have data. If the domain contained missing values, we would need to include **na.rm=TRUE** in the *svymean* function to be able to calculate statistics.

Functions *svymean* and *svyquantile* are then applied to calculate the estimated mean and quantiles of BMI and the proportion in each category of *bmi30*. The confidence intervals for quantiles differ slightly from those in SDA, which were calculated using SAS software under a slightly different algorithm. Adding **deff=TRUE** to *svymean* requests the design effect (deff) for each statistic.

```
# Request means and design effects
nhmeans<-svymean(~bmxbmi+bmi30, d0709sub, deff=TRUE)
degf(d0709sub)
## [1] 15
nhmeans
##                mean        SE    DEff
## bmxbmi  29.389101  0.253197  7.1248
## bmi300   0.607775  0.015856  5.7003
## bmi301   0.392225  0.015856  5.7003
confint(nhmeans,df=degf(d0709sub))
##                2.5 %      97.5 %
## bmxbmi  28.8494243  29.9287768
## bmi300   0.5739798   0.6415707
## bmi301   0.3584293   0.4260202
# Find quantiles
svyquantile(~bmxbmi, d0709sub, quantiles=c(0.05,0.25,0.5,0.75,0.95),
           ties = "rounded",ci=TRUE, interval.type="Wald")
## $quantiles
##              0.05      0.25       0.5      0.75      0.95
## bmxbmi  20.29893  24.35349  28.2349  33.06615  42.64092
##
## $CIs
## , , bmxbmi
##
##              0.05      0.25       0.5      0.75      0.95
## (lower  19.83403  23.92667  27.55465  32.35400  41.91584
## upper)  20.70609  24.84391  28.91359  33.64129  43.47766
```

The mean BMI for adults age 20 and over is 29.389 (using design object *d0709sub*), with 95% confidence interval [28.849, 29.929]. The estimated proportion of adults age 20 and over who have BMI > 30 (*bmi30*=1) is 0.392225 with 95% confidence interval [0.3584293, 0.4260202]. The confidence interval is calculated using a *t* distribution with 15 degrees of freedom (number of psus minus number of strata).

## 7.4    Univariate Plots from Complex Surveys

The *svyhist* and *svyboxplot* functions, along with *svydesign*, produce histograms and boxplots that incorporate the weights, as described in Chapter 7 of SDA.

**Examples 7.10, 7.11, and 7.12 of SDA.** These examples consider data in *htstrat*, a disproportional stratified sample of 160 women and 40 men.

**Histograms and smoothed density estimates.** Figure 7.1 shows the difference between a histogram constructed without the weights (left panel) and one constructed with the weights (right panel). Each histogram is overlaid with a smoothed density estimate.

```
data(htstrat)
# set graphics parameters, 1*2 plots, axis labels horizontal
par(mfrow=c(1,2),las=1,mar=c(2.1,4.1,2.1,0.3))
# Histogram overlaid with kernel density curve (without weight information)
# Displays the sample values, but does not estimate population histogram
# freq=FALSE changes the vertical axis to density
# breaks tell how many breakpoints to use
hist(htstrat$height,main="Without weights", xlab = "Height (cm)",
     breaks = 10, col="gray90", freq=FALSE, xlim=c(140,200), ylim=c(0,0.045))
# overlaid with kernel density curve
lines(density(htstrat$height),lty=1,lwd=2)

# Histogram (with weight information)
# create survey design object, weights calculated from fpc here
d0710 <- svydesign(id = ~1, strata = ~gender, fpc = c(rep(1000,160),rep(1000,40)),
               data = htstrat)
d0710
## Stratified Independent Sampling design
## svydesign(id = ~1, strata = ~gender, fpc = c(rep(1000, 160),
##     rep(1000, 40)), data = htstrat)
svyhist(~height,d0710, main="With weights",xlab = "Height (cm)",
        breaks = 10, col="gray90", freq=FALSE,xlim=c(140,200), ylim=c(0,0.045))
dens1<-svysmooth(~height,d0710,bandwidth=5)
lines(dens1,lwd=2) # draw the density line
```

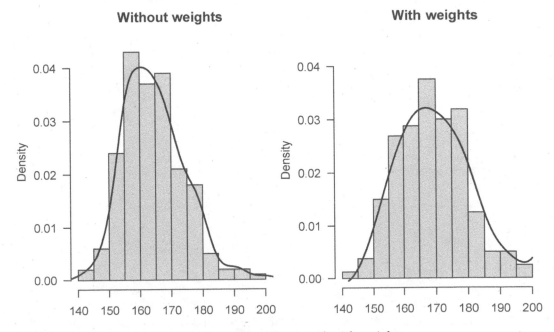

**FIGURE 7.1:** Histograms constructed without and with weights.

We call *svydesign* and *svyhist* to create a histogram that uses the survey information. The only difference between *svyhist* and *hist* is that *svyhist* includes the design object so that the histogram accounts for the weights.

```
svyhist(~height, design object, main=" ", xlab = " ", ylab= "")
```

In *hist* and *svyhist*, the *breaks* argument controls the number of bins, and `freq=FALSE` changes the vertical axis to density instead of frequency. The *density* function calculates a kernel density curve for unweighted data, and *svysmooth* performs density estimation using the weighted data. We allowed the bandwidth to be chosen automatically by the *density* function and specified a bandwidth of 5 in the *svysmooth* function (if you omit the *bandwidth* argument, it will be chosen automatically).

In Figure 7.1, the unweighted plot on the left displays the values in the sample but, because this is a disproportionally allocated sample, it does not estimate the histogram that would be obtained if we measured everyone in the population. The distribution appears skewed, reflecting the underrepresentation of men (who have greater average height) in the sample. The histogram and density estimate in the plot on the right incorporate the survey weights and thus can be interpreted as estimates of the histogram and density that would be obtained if the entire population were measured.

**Boxplots.** The *svyboxplot* function creates boxplots that include the weights; it is the survey analog of the *boxplot* function. Figure 7.2 displays boxplots for the full sample and separately for each gender.

```
par(mfrow=c(1,2),las=1,mar=c(2.1,4.1,2.1,0.3))
# boxplot (with weight information)
svyboxplot(height~1,d0710,ylab="Height",xlab=" ", main="Full sample")
svyboxplot(height~gender,d0710,ylab="Height",xlab="Gender",
          main="Separately by gender")
```

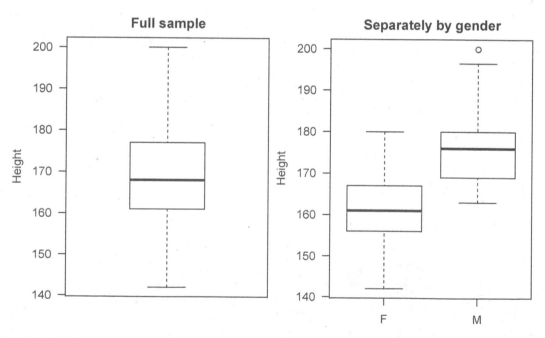

**FIGURE 7.2:** Boxplots of height for the full sample, and separately by gender.

**Histograms for domains.** What if you want to draw a histogram for just one domain from a complex survey? You can do that by using the *subset* function to redefine the object from *svydesign*. Figure 7.3 gives a histogram of BMI for adults age 20 and over, using the survey weights in the *nhanes* data. The smoothed density function is superimposed.

```
# Restore graphics settings
par(mfrow=c(1,1),las=1,mar=c(5.1, 4.1, 4.1, 2.1))
svyhist(~bmxbmi,d0709sub, main="Histogram of body mass index for adults",
        breaks = 30, col="gray90",xlab = "Body Mass Index (kg/m^2)")
dens2<-svysmooth(~bmxbmi,d0709sub)
lines(dens2,lwd=2)
```

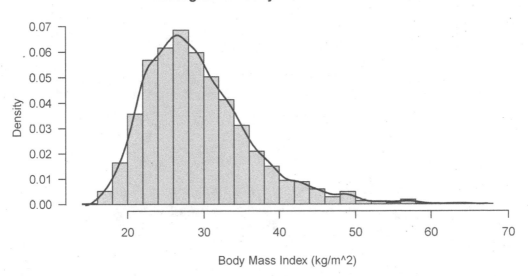

**FIGURE 7.3:** Histogram and density estimate for adults age 20 and over.

## 7.5 Scatterplots from Complex Surveys

Most of the plot types described in Section 7.6 of SDA can be drawn using functions *plot*, *svyplot*, *svyboxplot*, or *svycdf*.

**Unweighted plots.** The R functions *plot*, *hist*, and *boxplot* draw scatterplots, histograms, boxplots, and more. They do not incorporate the survey weights. In general, extra data preparation is needed to persuade these functions to draw graphs that estimate the population.

In some instances, however, you may want to examine the unweighted data. You may want to see how the weights affect the regression relationship between $x$ and $y$, or to identify unusual observations in the data. For that reason, we include an unweighted scatterplot of the NHANES data.

The following shows the code to produce the unweighted scatterplot of $y$ variable body mass index (*bmxbmi*) versus $x$ variable age (*ridageyr*) shown in Figure 7.4. This plot is for all ages, not just the subset of adults for which we computed summary statistics in Section 7.3. Numerous options are available for customizing the plot; we chose to use the default setting type="p" for the points and used plotting symbol '+' (pch=3), making the symbols small so more of them are displayed on the plot (cex=0.5). The optional *xlab* and *ylab* arguments allow customizing the axis labels. You can also set the minimum and maximum values for each axis using *xlim* and *ylim*.

```
# scatterplot without weights
par(las=1) # make tick mark labels horizontal
plot(nhanes$ridageyr,nhanes$bmxbmi,xlab="Age (years)",ylab="Body Mass Index",
    main="Scatterplot without weights",pch=3,cex=0.5,
    ylim=c(10,70),xlim=c(0,80))
```

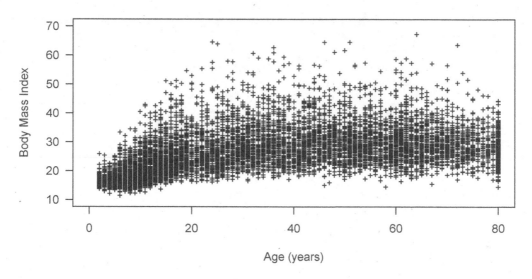

**FIGURE 7.4:** Scatterplot of BMI versus age (unweighted).

The unweighted scatterplot shows the relationship between $x$ and $y$ in the sample. If the sample is self-weighting, this scatterplot also estimates the relationship between $x$ and $y$ in the population. If the sample is not self-weighting, however, the relationship between $x$ and $y$ in the population may be different, and one of the other plots in this section should be used if you want to visualize the relationship between $x$ and $y$ in the population.

Although we do not recommend unweighted plots for non-self-weighting samples in general, sometimes, for small data sets, they are useful for seeing whether the relationship between $x$ and $y$ is the same with and without the weights (see Section 11.4 of SDA). The NHANES data, however, have so many data points that it is difficult to see patterns from the unweighted scatterplot in Figure 7.4.

**Plot subsample of data.** We can also select a subsample of the data that is approximately self-weighting, and draw the scatterplot of that subsample using the *plot* function. The plot of the subsample, displayed in Figure 7.5, then estimates the scatterplot that would be drawn from the population.

This subsample was selected with probability proportional to the weights and with replacement, using the *sample* function. Any method that will select a sample with probability proportional to the weights can be used, however, including the *UP* functions described in Table 6.1. If desired, you can draw multiple plots with different subsamples.

```
# select subsample with probability proportional to weights
set.seed(2847654)
subsamp<-sample(1:nrow(nhanes),500,replace=TRUE,prob=nhanes$wtmec2yr)
par(las=1) # make tick mark labels horizontal
plot(nhanes$ridageyr[subsamp],nhanes$bmxbmi[subsamp],
    xlab="Age (years)",ylab="Body Mass Index",
    main="Scatterplot of pps subsample",pch=3,cex=0.5,
    ylim=c(10,70),xlim=c(0,80))
```

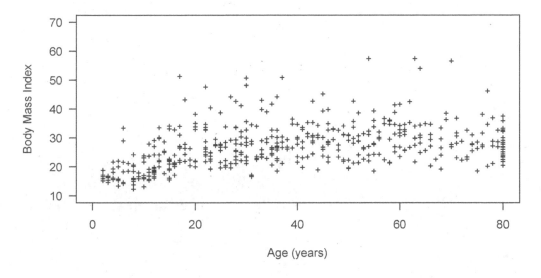

**FIGURE 7.5:** Scatterplot of BMI versus age (self-weighting subsample).

**Bubble plots.** If the sample is not self-weighting, a bubble plot displays the estimated shape of the population data. Bubble plots can be constructed using individual $(x, y)$ values, or using bins. We will show how to use *svyplot* to create bubble plots of the individual $(x, y)$ values.

The *svyplot* function is essentially the same as the *plot* function, except it adds the survey design object so that the weights are incorporated into the plots. You can use the same arguments to label the axes (*xlab, ylab*), set the extent of the plotting region (*xlim, ylim*), title the plot (*main*), and perform other formatting as in the *plot* function.

The *svyplot* function will construct a variety of types of plots that incorporate the weights, including bubble plots, "transparent" plots where the opacity of points is proportional to their weights, and binned hexagonal scatterplots.

Figure 7.6 shows a bubble plot for the NHANES data. The *inches* argument scales the bubbles; you may need to try several values until you find one that looks nice.

```
par(las=1) # make tick mark labels horizontal
svyplot(bmxbmi~ridageyr, design=d0709, style="bubble", inches=0.03,
        xlab="Age(years)",ylab="Body Mass Index",xlim=c(0,80),ylim=c(10,70),
        main="Weighted bubble plot of BMI versus age")
```

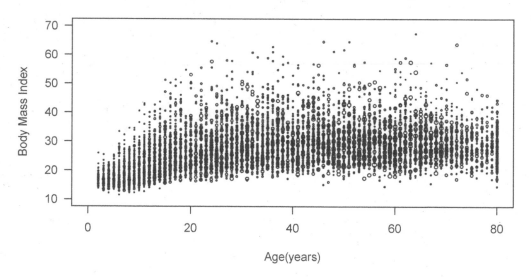

**FIGURE 7.6:** Bubble plot of BMI versus age (weighted).

**Plot data for a domain.** If we want to draw a plot for sample members in a specific domain, we can use the *subset* function to define a design object for the domain and then use *svyplot* with the subset design object.

Figure 7.7 shows a bubble plot of BMI versus age for the domain of non-Hispanic Asian Americans (having *ridreth3*=6).

```
# define subset
d0709subA<-subset(d0709, ridreth3==6)
par(las=1) # make tick mark labels horizontal
svyplot(bmxbmi~ridageyr, design=d0709subA, style="bubble",inches = 0.03,
        xlab="Age(years)",ylab="Body Mass Index",xlim=c(0,80),ylim=c(10,70),
        main="Weighted bubble plot of BMI versus age for Asian Americans")
```

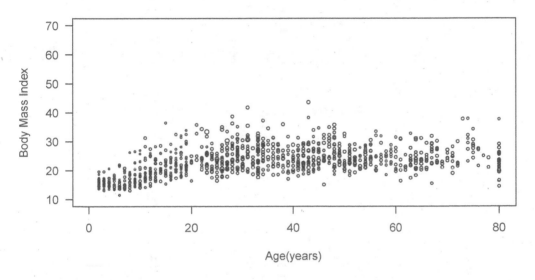

**FIGURE 7.7:** Bubble plot of BMI versus age for Asian Americans.

Note that for purposes of graphing data, it is also acceptable to consider just the subset of observations being graphed. This is because the scatterplots in this section incorporate the weights, but they do not use any other features of the survey design. Thus, if producing a subsampled graph for a domain that is similar to that in Figure 7.5, you could first create a subset of data consisting of the domain of interest, then select points with probabilities proportional to the weights, and then use the *plot* function to draw the scatterplot. With the *svyplot* function, however, it is easiest to create a subset design object for the domain being graphed.

**Side-by-side boxplots.** Boxplots will display the distributions of subgroups of the data. Figure 7.2 showed their use to display the distribution of height, separately for males and females.

They can also be used to display the bivariate relationship between two continuous variables. Simply partition the $x$ variable into a categorical variable that defines different ranges of the variable. You can use the *round* function to round the $x$ variable to the nearest multiple of a number. Alternatively, you can use the *cut* function to divide the range of $x$ into intervals with user-supplied cutpoints and code each value $x$ by the interval that contains it.

The easiest way to obtain side-by-side boxplots for survey data is through the *svyboxplot* function. Figure 7.8 shows boxplots of BMI by age groups that are formed by rounding the values of *ridageyr* to the nearest multiple of 5.

```
# include agegroup in the data frame
nhanes$agegroup<-5*round(nhanes$ridageyr/5)
d0709 <- svydesign(id = ~sdmvpsu, strata = ~ sdmvstra, nest=TRUE,
                weights=~wtmec2yr, data = nhanes)
par(las=1) # make tick mark labels horizontal
svyboxplot(bmxbmi~factor(agegroup),d0709,ylab="Body mass index",xlab="Age Group",
        ylim=c(10,70),main="Side-by-side boxplots of BMI for age groups")
```

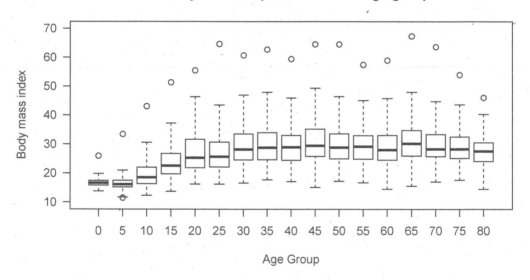

**FIGURE 7.8:** Boxplots of BMI by age groups.

**Smoothed trend line for mean.** The *svysmooth* function can carry out scatterplot smoothing and density estimation for weighted data. For the default method `"locpoly"`, the extra arguments are passed to *locpoly* from the `KernSmooth` package (Wand et al., 2020), which implements the smoothing methods described in Wand and Jones (1995). The default is local linear smoothing for the mean. If desired, the *bandwidth* argument can be included for a user-specified bandwidth.

Figure 7.9 shows the smoothed curve superimposed on the bubble plot in Figure 7.6. The color of the bubbles is changed to light gray for better visibility of the trend line.

```
# plot data bmxbmi~ridageyr
par(las=1) # make tick mark labels horizontal
svyplot(bmxbmi~ridageyr, design=d0709, style="bubble",basecol="gray",inches=0.03,
        xlab="Age(years)",ylab="Body Mass Index",xlim=c(0,80),ylim=c(10,70),
        main="Smoothed trend line with bubble plot of BMI versus age")
# plot smoothing trend line
# library(KernSmooth)  # install and load the package if not already done
smth<-svysmooth(bmxbmi~ridageyr,d0709)
lines(smth,lwd=2)
```

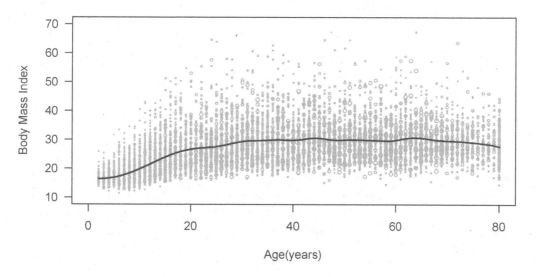

**FIGURE 7.9:** Smoothed trend line with bubble plot of BMI versus age.

Other smoothing methods may also be used to estimate trend lines. Figure 7.14 in SDA was created by calculating survey-weighted spline estimates (Zhang et al., 2015) of the trend line for a grid of $x$ points and then connecting the dots with the *lines* function.

**Smoothed trend lines for quantiles.** To fit smoothed lines for quantiles, change to method="quantreg" in *svysmooth*. This smooths the regression quantiles from package quantreg (Koenker et al., 2021). Instead of fitting smoothed lines to the means of different groups, separate lines are drawn that estimate each conditional quantile (Koenker, 2005) specified in the *quantile* argument. You can think of this as "connecting the dots" of the quantiles shown in the side-by-side boxplots in Figure 7.8. We requested the quantiles corresponding to probabilities 0.05, 0.25, 0.5, 0.75, and 0.95 with the argument taus=c(.05,.25,.5,.75,.95). This set of quantiles gives a good picture of the center (median) of the data, as well as the large and small values. Figure 7.10 includes the bubble plot and the smoothed line for each quantile.

```
# library(quantreg)) # install and load the package if not already done
# plot data bmxbmi~ridageyr
par(las=1) # make tick mark labels horizontal
svyplot(bmxbmi~ridageyr, design=d0709, style="bubble",basecol="gray",inches=0.03,
        xlab="Age (years)",ylab="Body Mass Index",xlim=c(0,80),ylim=c(10,70),
        main="Smoothed quantile trend lines")
# plot smoothed trend lines for quantiles
taus<-c(.05,.25,.5,.75,.95)
for (i in 1:length(taus)) {
  qsmth<-svysmooth(bmxbmi~ridageyr,d0709, quantile=taus[i],method="quantreg")
  lines(qsmth,lwd=1.2)
}
```

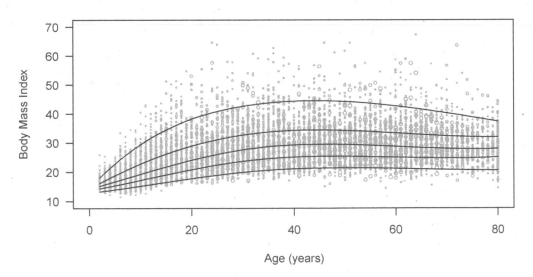

**FIGURE 7.10:** Smoothed quantile trend lines with bubble plot of BMI versus age.

**Customizing graphs.** Users who want to be able to do more customization of the graphs may want to write their own code. The graphs in SDA were produced with custom-written code (not using the *svyplot* function), and the file ch07.R includes code for using the ggplot2 graphics visualization package (Wickham et al., 2020) to create more types of scatterplots with survey data.

## 7.6  Additional Code for Exercises

Some of the exercises in Chapter 7 of SDA ask you to construct an empirical cumulative distribution function (ecdf) or an empirical probability mass function (epmf). Any population characteristic can be estimated using these functions. There are functions in R that calculate commonly requested statistics such as means, totals, quantiles, and regression coefficients from survey data, so the survey analyst typically does not calculate or graph the ecdf. For continuous variables, it is usually more informative to create a histogram or a smoothed density estimate than to view the epmf, which may contain a large number of spikes.

The ecdf and epmf are useful concepts for learning about how survey weights work, however, because they illustrate how the sample is used to create a reconstruction of the population. And both are easy to calculate in R.

**Example 7.5 of SDA.** Function *emppmf* from package `SDAResources` calculates the empirical probability mass function for a variable with associated weights. Call the function as

```
emppmf(y,sampling.weight)
```

to return a list with component vectors *vals*, the distinct values of $y$, and *epmf*, the value of the epmf corresponding to each $y$ value in *vals*.

Here we look at the stratified sample of heights in *htstrat*, where each female has a sampling weight of 1000/160 and each male has a sampling weight of 1000/40. It produces the estimated population proportion for each of the distinct values of *height*; the estimated proportion for value $y$ is the sum of the weights for observations having *height* = $y$ divided by the sum of all the weights in the sample. Figure 7.11 plots the epmf for *height*, using `type="h"` to draw vertical lines.

```
# Empirical pmf for stratified sample of heights
# define sampling weight
htstrat$sampwt <- 1000/sum(htstrat$gender=="F")
htstrat$sampwt[htstrat$gender=="M"] <- 1000/sum(htstrat$gender=="M")
# use function emppmf to calculate pmf
strresult <- emppmf(htstrat$height,htstrat$sampwt)
# plot
par(las=1)
plot(strresult$vals, strresult$epmf,type="h",xlab="Height Value, y (cm)",
     ylab="Empirical pdf",lwd=1.2,
     main="Empirical pdf for stratified sample of heights (weighted)")
```

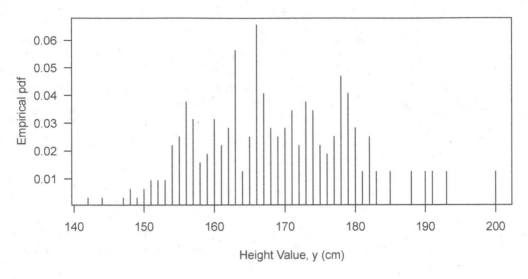

**FIGURE 7.11:** Empirical pmf for stratified sample of heights, using weights.

Function *ecdf* in base R will estimate the empirical cdf without weights, and function *svycdf* from the **survey** package will estimate the empirical cdf by incorporating survey weights. In contrast to *svyquantile*, *svycdf* does not interpolate but instead produces a right-continuous step function. Figure 7.12 shows the cdf of the population *htpop* (thick black) with the ecdf of sample *htstrat* without incorporating weights (medium red), and the ecdf of sample **htstrat** with weights (thin purple). We see that the medium red line is quite far off from the population cdf, while the thin purple line with weights is close to the population cdf.

```
# data(htstrat)
# Recall that
# d0710 <- svydesign(id = ~1, strata = ~gender, fpc = c(rep(1000,160),rep(1000,40)),
#                  data = htstrat)
cdf.weighted<-svycdf(~height, d0710)
cdf.weighted
## Weighted ECDFs: svycdf(~height, d0710)
## evaluate the function for height 144
cdf.weighted[[1]](144)
## [1] 0.00625
## compare to population and unweighted sample ecdfs.
cdf.pop<-ecdf(htpop$height)    # ecdf for population
cdf.samp<-ecdf(htstrat$height) # unweighted ecdf of sample
par(las=1,mar=c(5.1,4.1,2.1,2.1))
plot(cdf.pop, do.points = FALSE,
     xlab="Height value y",ylab="Empirical cdf",xlim=c(135,205),lwd=2.5,
     main="Empirical cdfs for population and sample")
lines(cdf.samp, col="red", do.points = FALSE, lwd=1.8)
lines(cdf.weighted[[1]], do.points = FALSE,  col ="purple",lwd=1)
legend("topleft", legend=c("Population", "Sample unweighted", "Sample weighted"),
       col=c("black", "red", "purple"),lwd=c(2.5,1.8,1),cex=0.8,bty="n")
```

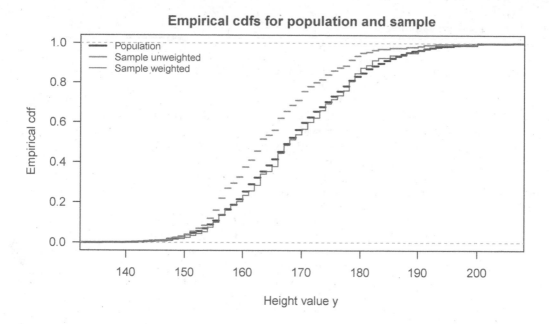

**FIGURE 7.12:** Empirical cdf of height for data *htpop*, and for data *htstrat* with and without weights.

## 7.7   Summary, Tips, and Warnings

Table 7.1 lists the major R functions used in this chapter.

**Tips and Warnings**

- For stratified multistage surveys, put the stratification variable in the *strata* argument of *svydesign*, and put the first-stage clustering variable in the *id* argument. Include the *weights* argument and do not include *fpc*. This will calculate the with-replacement variance approximations.

- If you want separate statistics for domains, first create the design object for the entire sample using the *svydesign* function, then use the *subset* or *svyby* function to calculate statistics for the domain of interest. This will ensure that standard errors for the domain statistics are calculated correctly. Do not subset the data first and then run *svydesign*— this can lead to incorrect standard error calculations.

- Incorporate the survey weights when constructing graphs if you want them to estimate the graphs that would be constructed if you had data from the entire population.

**TABLE 7.1**
Functions used for Chapter 7.

| Function | Package | Usage |
|---|---|---|
| sample | base | Select a with-replacement sample with unequal probabilities |
| subset | base | Work with a subset of a vector, matrix, or data frame |
| confint | stats | Calculate confidence intervals, add df for $t$ confidence interval |
| density | stats | Compute a kernel density estimate for self-weighting data |
| ecdf | stats | Calculate empirical cdf from self-weighting sample |
| par | graphics | Set graphics parameters |
| hist | graphics | Draw a histogram without weights |
| boxplot | graphics | Draw a boxplot without weights |
| plot | graphics | Draw a scatterplot without weights |
| mstage | sampling | Select a stratified multistage sample |
| svydesign | survey | Specify the survey design |
| svymean | survey | Calculate mean and standard error of mean |
| svyquantile | survey | Calculate quantiles and their confidence intervals |
| svyhist | survey | Draw a histogram of survey data, incorporating the weights |
| svyboxplot | survey | Draw boxplot of survey data, incorporating the weights |
| svyplot | survey | Draw scatterplot of survey data, incorporating the weights |
| svysmooth | survey | Estimate a smoothed density estimate or trend line from survey data |
| svycdf | survey | Calculate the empirical cdf from survey data |
| emppmf | SDAResources | Calculate the empirical probability mass function from survey data |

# 8

---

## *Nonresponse*

---

We have already seen most of the features needed to select survey samples and to compute estimates of means, totals, and quantiles using the sampling weights and survey design.

Nonresponse affects almost all surveys, however. Even samples of inanimate objects such as audit records may have missing accounts or missing items for accounts. This chapter looks at how R deals with missing data and points you to resources for calibrating survey data and performing imputations in R. The code is in file **ch08.R** on the book website.

---

## 8.1   How R Functions Treat Missing Data

We saw in Section 1.7 that R uses the value NA to denote missing data. If a data set uses a different code for missing values, such as $-99$, these must be converted to NA before analyzing the data so that R recognizes the values as missing. Otherwise, functions will treat these observations as though $y = -99$ and include them in calculations, sometimes with embarrassing results as when the estimated mean age is a negative number.

Data set **impute.csv** contains the data from Table 8.4 of SDA; it codes the missing values by $-99$. In the R data set *impute* in the **SDAResources** package, these missing values have been converted to NA. Variables *education*, *crime*, and *violcrime* all have missing values.

Most functions in R have a default method for how missing values are treated. The *mean* function, for example, returns NA if you attempt to find the mean of a vector that has missing values; if you want it to calculate the mean of the non-missing values, include argument **na.rm=TRUE**. The following shows how some of the base R functions and the *svymean* function treat the missing data in variables *crime* and *violcrime*, which take on values 1, 0, and NA.

```
data(impute)
impute$crime
##  [1]  0  1  0  1  1  0  1  0  0  0 NA  0  0  1  1  0  0  0  0 NA  0
is.na(impute$crime) # vector with TRUE for missing values
##  [1] FALSE FALSE FALSE FALSE FALSE FALSE FALSE FALSE FALSE  TRUE FALSE FALSE
## [13] FALSE FALSE FALSE FALSE FALSE FALSE  TRUE FALSE
# identify the rows with no missing values in columns 5-6
impute$cc<-complete.cases(impute[,5:6])
impute$cc
##  [1]  TRUE  TRUE  TRUE  TRUE  TRUE  TRUE FALSE  TRUE FALSE FALSE  TRUE  TRUE
## [13] FALSE  TRUE  TRUE  TRUE  TRUE  TRUE FALSE  TRUE
mean(impute$crime)  # returns NA
## [1] NA
mean(impute$crime,na.rm=TRUE) # calculates mean of non-missing values
## [1] 0.3333333
```

DOI: 10.1201/9781003228196-8

```
table(impute$crime,impute$violcrime) # excludes values missing in either variable
##
##      0  1
##  0  11  0
##  1   1  3
table(impute$crime,impute$violcrime,useNA="ifany") # counts NAs as category in table
##
##         0  1 <NA>
##  0     11  0    1
##  1      1  3    2
##  <NA>   1  0    1
# input design information, use relative weights of 1 for comparison with above
dimpute <- svydesign(id = ~1, weights = rep(1,20), data = impute)
dimpute
## Independent Sampling design (with replacement)
## svydesign(id = ~1, weights = rep(1, 20), data = impute)
# calculate survey mean and se
svymean(~crime,dimpute) # returns NA
##       mean SE
## crime   NA NA
svymean(~crime, dimpute, na.rm=TRUE)
##         mean     SE
## crime 0.33333 0.114
svytable(~violcrime+crime,dimpute)
##          crime
## violcrime  0  1
##         0 11  1
##         1  0  3
```

When **na.rm=TRUE** is included, the *mean* and *svymean* functions estimate the mean using the observations having non-missing values.

In a complex survey, however, omitting the missing values can affect the standard error calculations. A better option, for complex surveys where the missing values are not imputed, is to use the *subset* function to define the subset of observations with non-missing values on which the analysis is to be performed—treating the non-missing values as a domain, as seen in Section 7.3. Remember, though, that an analysis of complete cases alone can be subject to nonresponse bias. You may want to explore the sensitivity of your estimates to the assumptions about the missing values.

## 8.2   Poststratification and Raking

Unit nonresponse occurs when a unit selected for the sample provides no data. In some data sets, the nonrespondent unit may be represented in the data set, but with missing values for all survey responses. In others, nonrespondent units are missing entirely from the data set.

The most common method for trying to compensate for potential effects of nonresponse is to weight the data. The sampling frame may contain information that can be used in weighting class adjustments, or information known about the population from an external source may be used to poststratify or rake the data as described in Section 8.6 of SDA.

We have already seen the *postStratify* function from the **survey** package (Lumley, 2020) in Chapter 4. The package also has a function to rake the weights to marginal counts in a table. It takes the form

```
rake(design.object, sample.margins, population.margins)
```

where *sample.margins* is a list of formulas describing the sample margins and *population.margins* is a list giving the population counts for each raking variable.

Here is a simple example, using the data in Section 8.6.2 of SDA. The entries in the data frame *rakewtsum* are the sums of weights for persons in the sample falling into each cross-classification of gender by race. For this example, we assume that the data are from a sample of size 500 where each person has weight 6 and construct data frame *rakedf* that has 500 records listing the race and gender of each person (in most applications of raking, you will already have the data frame and will not need this step).

Function *rake* is then used to do raking adjustment on design object *drake* according to the population marginal totals of gender (*pop.gender*) and race (*pop.race*).

```
rakewtsum <- data.frame(gender=rep(c("F","M"),each=5),
                   race=rep(c("Black","White","Asian","NatAm","Other"),times=2),
                   wtsum=c(300,1200,60,30,30,150,1080,90,30,30))
rakewtsum # check data entry
##    gender  race wtsum
## 1       F Black   300
## 2       F White  1200
## 3       F Asian    60
## 4       F NatAm    30
## 5       F Other    30
## 6       M Black   150
## 7       M White  1080
## 8       M Asian    90
## 9       M NatAm    30
## 10      M Other    30
# Need data frame with individual records to use rake function
rakedf <- rakewtsum[rep(row.names(rakewtsum), rakewtsum[,3]/6), 1:2]
dim(rakedf)
## [1] 500   2
rakedf$wt <- rep(6,nrow(rakedf))
# Create the survey design object
drake <- svydesign(id=~1, weights=~wt, data=rakedf)
# Create data frames containing the marginal counts
pop.gender <- data.frame(gender=c("F","M"), Freq=c(1510,1490))
pop.race <- data.frame(race=c("Black","White","Asian","NatAm","Other"),
                   Freq=c(600,2120,150,100,30))
# Now create survey design object with raked weights
drake2 <- rake(drake, list(~gender,~race), list(pop.gender, pop.race))
drake2 # describes SRS with replacement
## Independent Sampling design (with replacement)
## rake(drake, list(~gender, ~race), list(pop.gender, pop.race))
# Look at first 10 entries in vector of raked weights
weights(drake2)[1:10]
##        1        1.1      1.2      1.3      1.4      1.5      1.6      1.7
## 7.511886 7.511886 7.511886 7.511886 7.511886 7.511886 7.511886 7.511886
##      1.8      1.9
## 7.511886 7.511886
```

```
# Look at sum of raked weights for raking cells
svytable(~gender+race, drake2)
##        race
## gender      Asian      Black      NatAm      Other      White
##      F   53.71714  375.59431   45.55940   13.66782 1021.46870
##      M   96.28286  224.40569   54.44060   16.33218 1098.53130
# Look at sum of raked weights for margins
svytotal(~factor(gender),drake2)
##                   total      SE
## factor(gender)F    1510  0.0028
## factor(gender)M    1490  0.0028
svytotal(~factor(race),drake2)
##                  total SE
## factor(race)Asian   150  0
## factor(race)Black   600  0
## factor(race)NatAm   100  0
## factor(race)Other    30  0
## factor(race)White  2120  0
```

After raking, the standard errors for the population counts for the raking categories equal 0 (approximately 0 for gender since the raking procedure stops when a preset tolerance is reached; you can change that tolerance if desired). After raking, there is no longer any sampling variability for the estimated population counts for each race category—the raking process has forced them to equal the population counts from the external source, which are assumed to be known exactly.

You can also see the raking iterations, if desired, by typing

```
rake(drake, list(~gender,~race), list(pop.gender, pop.race),
     control=list(verbose=TRUE)).
```

**Other R functions for nonresponse weight adjustments.** The `survey` package has several other functions for nonresponse weight adjustments. The *calibrate* function will perform calibration with general auxiliary information, as described in Section 11.6 of SDA. The *trimWeights* function will trim weights that are outside of user-defined bounds for the weights, and redistribute the trimmings among the untrimmed observations (this keeps the sum of the weights the same). Valliant et al. (2018) go through the detailed steps for constructing weights and describe additional resources for nonresponse weight adjustments using R.

## 8.3  Imputation

Many data sets have item nonresponse in addition to unit nonresponse. Item nonresponse occurs when a unit has responses for some of the items on the survey but missing values for other items. A person may answer questions about age, race, gender, and health outcomes but decline to answer questions about income. Or the data editing process may uncover logical inconsistencies in the data, such as a 6-year-old who is listed as being married. If the discrepancy cannot be resolved, both age and marital status may be recoded as missing for that person. In this section, we assume that the imputation is limited to missing items.

**Example 8.9 of SDA.** It is easy to perform simple imputation methods such as cell mean or regression imputation in R. We do not recommend these methods for most applications, since they do not preserve multivariate relationships and do not provide a means of accounting for the imputation when calculating standard errors. But they may be useful when there are, say, a handful of missing items in one of the weighting variables or if it is desired to have an initial imputation before moving on to one of the more advanced imputation methods.

Cell mean imputation is a special case of regression imputation, where the explanatory variables are the factors defining the cells. For cell mean imputation, we first define the cells that cross-classify the observations by gender and age group, then use the *tapply* function to calculate the mean age for the non-missing values in each cell. Note that we also calculate a matrix *impute.flag* that tells which observations were imputed.

```
##### cell mean imputation #####
impute$education
## [1] 16 NA 11 NA 12 NA 20 12 13 10 12 12 11 16 14 11 14 10 12 10
impute.cm<-impute
# define matrix giving imputation flags, TRUE for each missing value
impute.flag<-is.na(impute)
# fit two-way model with interaction, omit NAs from model-fitting
edmodel<-lm(education~factor(gender)*factor(age>=35),
          data=impute.cm,na.action=na.omit)
# replace missing values with imputations from model
newdata<- impute[is.na(impute$education),]
impute.cm$education[is.na(impute$education)] <- predict(edmodel,newdata)
impute.cm$education
## [1] 16.00 12.25 11.00 11.25 12.00 12.25 20.00 12.00 13.00 10.00 12.00 12.00
## [13] 11.00 16.00 14.00 11.00 14.00 10.00 12.00 10.00
```

The other variables can be imputed similarly. Although this simple method might suffice if there are only a handful of missing items to be imputed, it will not preserve multivariate relationships. If you want to perform multivariate analyses of imputed data, the imputation model must include the relationships you will be studying in the analysis model. This requires a much more complicated procedure than a simple cell mean or regression model. The imputer must decide on an imputation method and model(s), and decide how to incorporate survey design features into the imputation (see Little and Vartivarian, 2003; Reiter et al., 2006; Kott, 2012, for discussions of this issue).

For most applications, we recommend using one of the many contributed packages for R that perform imputation for survey data. We briefly describe four of them here; Yadav and Roychoudhury (2018) review additional packages.

- The `Hmisc` package (Harrell, 2021) contains functions for single and multiple imputation using additive regression and predictive mean matching.

- The `VIM` package (Kowarik and Templ, 2016; Templ et al., 2021) performs model-based, hot-deck, and nearest-neighbor imputations. The package also provides tools for exploring the missing data patterns and imputations through visualization.

- The `FHDI` package (Im et al., 2018; Cho et al., 2020) imputes multivariate missing data using the fractional hot deck imputation method described by Kim and Fuller (2004) and Kim (2011). In this method, multiple donors contribute to each imputation.

- The `mice` package (van Buuren et al., 2021) performs multivariate imputation by chained equations. This method relies on a sequence of regression models that predict the missing values for each response variable in turn. This is the package that we usually use when imputing data. Azur et al. (2011) and van Buuren (2018) describe how to perform imputations using this method.

## 8.4   Summary, Tips, and Warnings

The `survey` package has functions for raking and poststratification that can be used to perform simple nonresponse adjustments of the weights.

Table 8.1 lists the major R functions used in this chapter.

**TABLE 8.1**
Functions used for Chapter 8.

| Function | Package | Usage |
|----------|---------|-------|
| is.na | base | Indicate which values are missing; the function returns TRUE if the value is missing and FALSE otherwise |
| subset | base | Work with a subset of a vector, matrix, or data frame |
| complete.cases | stats | Indicate which records have complete data; the function returns a vector with the value TRUE if the record has no missing values and FALSE if at least one item is missing |
| lm | stats | Fit a linear model to a data set (not using survey methods) |
| predict | stats | Obtain predicted values from a model object |
| svydesign | survey | Specify the survey design |
| svymean | survey | Calculate mean and standard error of mean |
| svytotal | survey | Calculate total and standard error of total |
| postStratify | survey | Adjust the sampling weights using poststratification |
| rake | survey | Carry out poststratification to table margins using raking |

**Tips and Warnings**

- Check how missing values are coded before analyzing your data set, and recode the missing values to NA.

- The functions *postStratify* and *rake* in the `survey` package will perform poststratification adjustments to the weights.

- If the survey has item nonresponse and imputation is not used to fill in missing values, treat the observations with non-missing values as a domain using the *subset* function.

- Several R packages are available that perform imputation. Be aware, though, that imputed values are only as good as the model that produces them. Performing a good imputation requires a lot of expertise; the references in the For Further Reading section of Chapter 8 of SDA can help you get started. See Haziza (2009) for a summary of approaches.

# 9

## Variance Estimation in Complex Surveys

In this chapter, we discuss variance estimation in complex surveys. We have already seen the use of linearization methods to calculate variances in functions such as *svymean* and *svytotal* in the **survey** package (Lumley, 2020). This chapter will focus on the other methods for calculating variances—random groups, balanced repeated replication (BRR), jackknife, and bootstrap. The code is in file **ch09.R** on the book website.

### 9.1 Replicate Samples and Random Groups

The random group methods in Section 9.2 of SDA are seldom used in practice. But they are of interest because they motivate how replication methods work in general, and they can be useful for providing quick-and-easy variance estimates for a complex survey design. The methods require three steps:

1. Select the replicate samples from the population, or divide the probability sample among the random groups.

2. Calculate the statistic of interest from each replicate, using the survey weights.

3. Use the estimated statistics from the replicates to calculate the standard error.

**Example 9.3 of SDA.** In this example, we consider estimating the variance of the ratio of out-of-state tuition fee (*tuitionfee_ out*) and in-state tuition fee (*tuitionfee_ in*) for a population of public colleges using replicate samples. Data *public_ college*, created below, consists of 500 public colleges and universities from the *college* data. This data set serves as the population from which we draw replicate samples.

For Step 1, we use a loop and the *srswor* function (package **sampling**, Tillé and Matei, 2021) to select five independent replicate SRSs, each of size 10, from *public_ college* (one could also write a function to do this). We print the values of the selected variables for the colleges in the fifth replicate sample. Note that the weights for each replicate sample sum to the population size, 500. The replicate samples in this book differ from those in SDA, which were selected using SAS software.

For Step 2, compute the statistic of interest from each replicate sample. We calculate the ratio of average out-of-state tuition $ybar[i]$ to average in-state tuition $xbar[i]$ for each replicate sample $i$ by calling the *svymean* function.

Variable *thetahat* contains the values of $\hat{\theta}_i$ for the five replicate samples. These values are the only part of the output used in Step 3; The standard errors given by the *svymean* function are ignored—only the point estimates of the means are used.

DOI: 10.1201/9781003228196-9

For Step 3, treat the 5 estimated ratios in *thetahat* as independent and identically distributed observations, and calculate their mean $\tilde{\theta}$. We also calculate a 95% confidence interval, which can be done either by formula or with the *t.test* function.

```
data(college)
# define population with public colleges and universities
public_college<-college[college$control==1,]
N<-nrow(public_college) #500
# select five SRSs and calculate means
xbar<-rep(NA,5)
ybar<-rep(NA,5)
set.seed(8126834)
for(i in 1:5){
  index <- srswor(10,N)
  replicate <- public_college[(1:N)[index==1],]
  # save replicate in a data frame if you want to keep it for later analyses
  # define design object (since SRS, weights are computed from fpc)
  dcollege<-svydesign(id = ~1, fpc = ~rep(500,10), data = replicate)
  # calculate mean of in-state and out-of-state tuition fees
  xbar[i]<-coef(svymean(~tuitionfee_in, dcollege))
  ybar[i]<-coef(svymean(~tuitionfee_out,dcollege))
}
# print the 5th replicate sample
replicate[,c(2,24:25)]
##                                     instnm tuitionfee_in tuitionfee_out
## 459                    Coppin State University          8873          15144
## 474                          Towson University          9940          23208
## 556              University of Michigan-Flint         11304          22065
## 674                 University of Nevada-Reno          7599          22236
## 735                      CUNY Brooklyn College          7240          14910
## 853   University of North Carolina at Greensboro       7331          22490
## 1024    Millersville University of Pennsylvania      12226          22196
## 1030   Pennsylvania State University-Main Campus     18454          34858
## 1359           Texas A&M University-San Antonio        8656          21159
## 1368        University of North Texas at Dallas        9139          21589
# calculate and print the five ratio estimates
thetahat<-ybar/xbar
thetahat
## [1] 2.172545 2.055528 2.107828 2.213799 2.181924
# calculate mean of the five ratio estimates, and SE
thetatilde<-mean(thetahat)
thetatilde
## [1] 2.146325
setheta<-sqrt(var(thetahat)/5)
# calculate confidence interval by direct formula using t distribution
c( thetatilde- qt(.975, 4)*setheta, thetatilde+ qt(.975, 4)*setheta)
## [1] 2.067224 2.225426
# easier: use t.test function to calculate mean and confidence interval
t.test(thetahat)
##
##   One Sample t-test
##
## data:  thetahat
## t = 75.336, df = 4, p-value = 1.861e-07
## alternative hypothesis: true mean is not equal to 0
```

```
## 95 percent confidence interval:
##   2.067224 2.225426
## sample estimates:
## mean of x
##   2.146325
```

The output shows that the mean of the ratios estimated from the five replicate samples is 2.146 and the 95% confidence interval for the population ratio is $[2.067, 2.225]$. The 95% confidence interval is calculated as $\tilde{\theta} \pm t \, \mathrm{SE}(\tilde{\theta})$, where $t$ is the critical value from a $t$ distribution with 4 (number of replicates minus one) degrees of freedom (df). This critical value is 2.78, giving a wider interval than would be obtained from a variance estimate with more df.

In Example 9.3, the survey weights are the only design feature used for the calculations. Even if a replicate sample has stratification or clustering, that information is not needed to calculate the point estimate of the parameter of interest $\theta$ for the replicates. The effect of the stratification or clustering on the variance is incorporated in the variability among the $\hat{\theta}_i$'s. For example, if clustering decreases the precision for $\hat{\theta}_i$, then the estimates $\hat{\theta}_i$ will vary more from replicate to replicate, and the decreased precision will show up in a large value of the sample variance for the replicate values $\hat{\theta}_i$.

We illustrated the method for estimating a ratio, but the same method can be used for any statistic you would like to estimate. The method can also be applied to statistics that are not calculated by the survey analysis functions. All you need to do is to calculate the statistic of interest for each replicate using the survey weights, then apply $t$ confidence interval methods to the statistics calculated from the replicates.

**Random groups.**  After dividing the survey data into $R$ random groups, the procedure for calculating the estimator and variance is exactly the same as Steps 1 to 3 for replicate sample methods. Note that if you are estimating population totals, you need to scale the weights for each random group so they sum to the population size.

**Example 9.4 of SDA.** This example illustrates the random group method with the Survey of Youth in Custody (*syc*) data. The whole sample is divided into seven random groups by variable *randgrp*. The *svyby* function computes the mean of variable *age* using *svymean* separately for each random group. Each is calculated using the survey weight *finalwt*.

Only the point estimates for *age*, printed in *repmean*, are needed to calculate the mean and confidence interval. The standard errors are not used. Thus, although the survey has stratification and clustering, we do not need to include that information in the *svydesign* function when calculating variances with the random groups method—the estimated means of each group depend only on the weights in *finalwt*.

```
data(syc)
dsyc<-svydesign(id = ~1, weights = ~finalwt, data = syc)
repmean<-svyby(~age, ~randgrp, dsyc, svymean)
repmean # we use only the means, not the SEs
##   randgrp      age         se
## 1       1 16.54947 0.1171541
## 2       2 16.66331 0.1133751
## 3       3 16.82544 0.1242695
## 4       4 16.05688 0.1240046
## 5       5 16.31776 0.1160307
## 6       6 17.02798 0.1181861
## 7       7 17.26605 0.1110258
```

Let $\hat{\theta}_r$ be the estimate from the $r$th replicate sample, $r = 1, 2, \cdots, 7$, $\tilde{\theta}$ be the mean of the seven random group means, and $\hat{\theta}$ be the estimated mean age using the whole data set *syc*. We can calculate two variance estimates: *SEthetatilde* calculates the square root of

$$\hat{V}_1(\tilde{\theta}) = \frac{1}{R}\frac{1}{R-1}\sum_{r=1}^{R}\left(\hat{\theta}_r - \tilde{\theta}\right)^2 \tag{9.1}$$

and *SEthetahat* calculates the square root of

$$\hat{V}_2(\hat{\theta}) = \frac{1}{R}\frac{1}{R-1}\sum_{r=1}^{R}\left(\hat{\theta}_r - \hat{\theta}\right)^2. \tag{9.2}$$

```
# Estimate and SE 1 (could also use t.test function)
thetatilde<-mean(repmean$age)
SEthetatilde<- sqrt( (1/7)*var(repmean$age) )
# Estimate and SE 2
thetahat<-coef(svymean(~age,dsyc))
SEthetahat<- sqrt((1/7)*(1/6)*sum((repmean$age-thetahat)^2))

#calculate confidence interval by direct formula using t distribution
Mean_CI1 <- c(thetatilde, SEthetatilde, thetatilde- qt(.975, 7-1)*SEthetatilde,
              thetatilde+ qt(.975, 7-1)*SEthetatilde)
names(Mean_CI1) <- c("thetatilde","SE","lower CL", "upper CL")
Mean_CI1
## thetatilde          SE   lower CL   upper CL
## 16.6724103   0.1559995 16.2906932 17.0541274
Mean_CI2 <- c(thetahat,SEthetahat, thetahat- qt(.975, 7-1)*SEthetahat,
              thetahat+ qt(.975, 7-1)*SEthetahat)
names(Mean_CI2) <- c("thetahat","SE","lower CL", "upper CL")
Mean_CI2
##   thetahat          SE   lower CL   upper CL
## 16.6392931   0.1565843 16.2561452 17.0224411
```

The estimated mean of age using *thetatilde* is 16.67241 with a confidence interval of $[16.29069, 17.05413]$. The estimated mean of age using *thetahat* is 16.63929 with a confidence interval of $[16.25615, 17.02244]$. A $t$ distribution with 6 (the number of random groups minus 1) df is used. Again, though, with few random groups, the confidence interval is wider than it would be if a variance estimation method having more df were used. We discuss such methods in the next section.

## 9.2   Constructing Replicate Weights

Replicate weights for balanced repeated replication (BRR), jackknife, and bootstrap methods can be created with the *as.svrepdesign* function in the **survey** package (Lumley, 2020). It creates a replicate-weights survey design object from a survey design object that contains stratification and clustering information.

The basic function call, shown here for BRR, is

```
as.svrepdesign(design.object, type = "BRR")
```

where *design.object* is a design object that has been created using *svydesign* (containing the stratification and clustering information) and *type* specifies the replication method to be used.

## 9.2.1 Balanced Repeated Replication

**Example 9.5 of SDA.** This example shows how to use BRR to calculate variances for the small data set in Table 9.2 of SDA. Here, we assume that $N = 10,000$. Variable *wt* contains the sampling weight, which is $N_h/2$ for this sample with $n_h = 2$ sampled observations per stratum and is obtained by $N(N_h/N)/n_h = (N)(strfrac)/2$. We request type="BRR" in the *as.svrepdesign* function.

The function creates 8 replicate weights because that is the smallest multiple of 4 that is larger than 7, the number of strata.

```
brrex<-data.frame(strat = c(1,1,2,2,3,3,4,4,5,5,6,6,7,7),
        strfrac =c(0.3,0.3,0.1,0.1,0.05,0.05,0.1,0.1,0.2,0.2,0.05,0.05,0.2,0.2),
        y =c(2000,1792,4525,4735,9550,14060,800,1250,9300,7264,13286,12840,2106,2070)
        )

brrex$wt <- 10000*brrex$strfrac/2
brrex
##    strat strfrac     y    wt
## 1      1    0.30  2000  1500
## 2      1    0.30  1792  1500
## 3      2    0.10  4525   500
## 4      2    0.10  4735   500
## 5      3    0.05  9550   250
## 6      3    0.05 14060   250
## 7      4    0.10   800   500
## 8      4    0.10  1250   500
## 9      5    0.20  9300  1000
## 10     5    0.20  7264  1000
## 11     6    0.05 13286   250
## 12     6    0.05 12840   250
## 13     7    0.20  2106  1000
## 14     7    0.20  2070  1000
dbrrex<-svydesign(id=~1, strata=~strat,weights=~wt,data=brrex)
dbrrex # stratified random sample
## Stratified Independent Sampling design (with replacement)
## svydesign(id = ~1, strata = ~strat, weights = ~wt, data = brrex)
# convert to BRR replicate weights
dbrrexbrr <- as.svrepdesign(dbrrex, type="BRR")
dbrrexbrr # identifies as BRR
## Call: as.svrepdesign(dbrrex, type = "BRR")
## Balanced Repeated Replicates with 8 replicates.
# now use the replicate weights to calculate the mean and confidence interval
svymean(~y,dbrrexbrr)
##      mean      SE
## y 4451.7 236.42
degf(dbrrexbrr)
## [1] 7
confint(svymean(~y,dbrrexbrr),df=7)
##      2.5 %   97.5 %
## y 3892.664 5010.736
```

The estimated mean value of $y$ is 4451.7 with a standard error of 236.42, and a confidence interval of [3892.664, 5010.736].

You can also look at the replicate weights multiplier if desired. Note that the two observations in each stratum have complementary patterns.

```
# can look at replicate weight multiplier if desired
# this is to illustrate weight structure; it is not needed for data analysis
dbrrexbrr$repweights$weights
##          [,1] [,2] [,3] [,4] [,5] [,6] [,7] [,8]
##   [1,]    2    0    2    0    2    0    2    0
##   [2,]    0    2    0    2    0    2    0    2
##   [3,]    2    2    0    0    2    2    0    0
##   [4,]    0    0    2    2    0    0    2    2
##   [5,]    2    0    0    2    2    0    0    2
##   [6,]    0    2    2    0    0    2    2    0
##   [7,]    2    2    2    2    0    0    0    0
##   [8,]    0    0    0    0    2    2    2    2
##   [9,]    2    0    2    0    0    2    0    2
##  [10,]    0    2    0    2    2    0    2    0
##  [11,]    2    2    0    0    0    0    2    2
##  [12,]    0    0    2    2    2    2    0    0
##  [13,]    2    0    0    2    0    2    2    0
##  [14,]    0    2    2    0    2    0    0    2
```

**Fay's method for BRR.** The *as.svrepdesign* function will also construct replicate weights using Fay's variation of BRR (Dippo et al., 1984; Judkins, 1990). Simply specify `type="Fay"` instead of `type="BRR"`. To set $\varepsilon$ for Fay's method equal to 0.5, add the argument `fay.rho = 0.5`. This results in weight multipliers 1.5 and 0.5 instead of the multipliers 0 and 2 for the BRR example above.

```
dbrrexfay <- as.svrepdesign(dbrrex, type="Fay",fay.rho=0.5)
svymean(~y,dbrrexfay)
##       mean      SE
## y 4451.7 236.42
confint(svymean(~y,dbrrexfay),df=7)
##        2.5 %   97.5 %
## y 3892.664 5010.736
# look at replicate weights for contrast with regular BRR
# note values for replicate weight multiplier are now 1.5 and 0.5
dbrrexfay$repweights$weights
##          [,1] [,2] [,3] [,4] [,5] [,6] [,7] [,8]
##   [1,]   1.5  0.5  1.5  0.5  1.5  0.5  1.5  0.5
##   [2,]   0.5  1.5  0.5  1.5  0.5  1.5  0.5  1.5
##   [3,]   1.5  1.5  0.5  0.5  1.5  1.5  0.5  0.5
##   [4,]   0.5  0.5  1.5  1.5  0.5  0.5  1.5  1.5
##   [5,]   1.5  0.5  0.5  1.5  1.5  0.5  0.5  1.5
##   [6,]   0.5  1.5  1.5  0.5  0.5  1.5  1.5  0.5
##   [7,]   1.5  1.5  1.5  1.5  0.5  0.5  0.5  0.5
##   [8,]   0.5  0.5  0.5  0.5  1.5  1.5  1.5  1.5
##   [9,]   1.5  0.5  1.5  0.5  0.5  1.5  0.5  1.5
##  [10,]   0.5  1.5  0.5  1.5  1.5  0.5  1.5  0.5
##  [11,]   1.5  1.5  0.5  0.5  0.5  0.5  1.5  1.5
##  [12,]   0.5  0.5  1.5  1.5  1.5  1.5  0.5  0.5
##  [13,]   1.5  0.5  0.5  1.5  0.5  1.5  1.5  0.5
##  [14,]   0.5  1.5  1.5  0.5  1.5  0.5  0.5  1.5
```

**Example 9.6 of SDA.** Now let's create BRR weights for a data set with stratification and clustering: the NHANES data. First, use *svydesign* to include primary sampling unit (psu) *sdmvpsu*, stratum *sdmvstra*, and weight *wtmec2yr* information, and then use the *as.svrepdesign* function to construct replicate weights using BRR by adding `type="BRR"`. By specifying the stratification and clustering information through *svydesign*, you ensure that observations in the same psu are kept together during the replicate weight construction.

This analysis finds summary statistics of body mass index (BMI, in variable *bmxbmi*) for adults age 20 and over, so the *subset* function is also used. Note that the df for the subset is the same as for the full sample—there are members from this domain in each psu, so there are (number of psus − number of strata) df.

```
data(nhanes)
nhanes$age20d<-rep(0,nrow(nhanes))
nhanes$age20d[nhanes$ridageyr >=20 & !is.na(nhanes$bmxbmi)]<-1
dnhanes<-svydesign(id=~sdmvpsu, strata=~sdmvstra,nest=TRUE,
                   weights=~wtmec2yr,data=nhanes)
dnhanesbrr <- as.svrepdesign(dnhanes, type="BRR")
# look at subset of adults age 20+
dnhanesbrrsub<-subset(dnhanesbrr, age20d =='1')
degf(dnhanes)
## [1] 15
degf(dnhanesbrrsub) # same df
## [1] 15
# calculate mean
bmimean<-svymean(~bmxbmi, dnhanesbrrsub)
bmimean
##           mean     SE
## bmxbmi 29.389  0.261
confint(bmimean,df=15)
##            2.5 %   97.5 %
## bmxbmi 28.83279 29.94541
# calculate quantiles
svyquantile(~bmxbmi, dnhanesbrrsub, quantiles=c(0.25,0.5,0.75,0.95),
            ties = "rounded")
## Statistic:
##           bmxbmi
## q0.25 24.35349
## q0.5  28.23490
## q0.75 33.06615
## q0.95 42.64092
## SE:
##           bmxbmi
## q0.25 0.2215986
## q0.5  0.3241246
## q0.75 0.3139102
## q0.95 0.3436826
```

The mean and quantiles are estimated using the full sample weights, and the replicate weights are used to calculate the standard errors. The estimated mean BMI for adults is 29.389 with a standard error of 0.261 and a 95% confidence interval of [28.833, 29.945]. You can also include optional argument `return.replicates=TRUE` in the *svymean* statement in case you want to look at the statistics calculated for each replicate.

Note that these replicate weights, constructed from the NHANES final weights available on the public-use data file, do not account for the effects of poststratification on the variances.

See Section 9.2.4 for how to calculate replicate weights so that the variance estimates include the effects of weighting adjustments.

**Domain estimates with replicate weights.**  We carried out this analysis on the subset of observations with *age20d*=1. The *subset* function is used exactly the same way with a replicate-weight design object as with a strata/cluster design object. Make sure, though, that you define the replicate-weight design object on the full data set first, and then use the *subset* function to define the domain. This ensures that the full design information is used to calculate standard errors.

## 9.2.2   Jackknife

The **survey** package will create replicate weights for two types of jackknife: JK1 and JKn. These are described in detail by Brick et al. (2000) and Chapter 15 of Valliant et al. (2018). Briefly, JKn is the jackknife for stratified multistage sampling described in Chapter 9 of SDA. If there are $n_h$ psus in stratum $h$ and observation $i$ has weight $w_i$, then the JKn jackknife weights are defined as follows:

$$w_{i(hj)} = \begin{cases} w_i & \text{if observation unit } i \text{ is not in stratum } h \\ 0 & \text{if observation unit } i \text{ is in psu } j \text{ of stratum } h \\ \dfrac{n_h}{n_h - 1} w_i & \text{if observation unit } i \text{ is in stratum } h \text{ but not in psu } j. \end{cases} \tag{9.3}$$

The weights $w_{i(hj)}$ are used to calculate $\hat{\theta}_{(hj)}$ for each replicate, and

$$\hat{V}_{\mathrm{JKn}}(\hat{\theta}) = \sum_{h=1}^{H} \frac{n_h - 1}{n_h} \sum_{j=1}^{n_h} \left( \hat{\theta}_{(hj)} - \hat{\theta} \right)^2. \tag{9.4}$$

Jackknife JK1 is a special case of JKn where there is one stratum.

**Jackknife weights for an SRS.**  Let's start by looking at how the *as.svrepdesign* function works using jackknife variance estimation for an SRS, then move on to complex sample designs.

**Example 9.7 of SDA.**  Data *collegerg* shows the values of in-state and out-of-state tuition for five replicate samples, each of size 10 (these samples were selected with SAS software, and differ from the replicate samples selected earlier in this chapter). We define data `collegerg1` to be replicate sample 1 (having *repgroup* = 1).

For comparison purposes, we first look at the linearization (Taylor series) variance estimate of the mean in-state tuition, mean out-of-state tuition, and ratio of mean out-of-state tuition to mean in-state tuition. Note that we did not include the *fpc* argument in *svydesign*, so the with-replacement variance is calculated.

```
data(collegerg)
collegerg1<-collegerg[collegerg$repgroup==1,]
collegerg1$sampwt<-rep(500/10,10)
# calculate SEs of means and ratio using linearization
dcollegerg1<-svydesign(id=~1, weights=~sampwt,data=collegerg1)
means.lin<-svymean(~tuitionfee_in+tuitionfee_out, dcollegerg1)
means.lin
##                    mean       SE
## tuitionfee_in    8913.3    454.46
## tuitionfee_out  21614.7   2325.15
```

```
confint(means.lin,df=degf(dcollegerg1))
##                   2.5 %    97.5 %
## tuitionfee_in    7885.247  9941.353
## tuitionfee_out 16354.843 26874.557
ratio.lin<-svyratio(~tuitionfee_out,~tuitionfee_in,dcollegerg1)
ratio.lin
## Ratio estimator: svyratio.survey.design2(~tuitionfee_out, ~tuitionfee_in,
##      dcollegerg1)
## Ratios=
##                tuitionfee_in
## tuitionfee_out      2.424994
## SEs=
##                tuitionfee_in
## tuitionfee_out     0.2311776
confint(ratio.lin,df=degf(dcollegerg1))
##                                 2.5 %    97.5 %
## tuitionfee_out/tuitionfee_in 1.902034 2.947954
```

Now let's calculate the jackknife variance by omitting observation $j$ in replicate $j$. We use function *as.svrepdesign* with `type="JK1"` to create jackknife weights for this SRS of size 10. We then use *svymean* and *svyratio* to calculate the standard errors of the estimated means and ratio with the jackknife weights.

```
## define jackknife replicate weights design object
dcollegerg1jk <- as.svrepdesign(dcollegerg1, type="JK1")
dcollegerg1jk
## Call: as.svrepdesign(dcollegerg1, type = "JK1")
## Unstratified cluster jacknife (JK1) with 10 replicates.
# now look at jackknife SE for means
# these are same as linearization since SRS and statistic = mean
svymean(~tuitionfee_in + tuitionfee_out, dcollegerg1jk)
##                   mean      SE
## tuitionfee_in    8913.3  454.46
## tuitionfee_out 21614.7 2325.15
# jackknife SE for ratio
svyratio(~tuitionfee_out, ~tuitionfee_in, design = dcollegerg1jk)
## Ratio estimator: svyratio.svyrep.design(~tuitionfee_out, ~tuitionfee_in,
##      design = dcollegerg1jk)
## Ratios=
##                tuitionfee_in
## tuitionfee_out      2.424994
## SEs=
##            [,1]
## [1,] 0.2314828
```

The above output shows the statistics produced by the *svymean* and *svyratio* functions using linearization and jackkinfe methods. Typing `dcollegerg1jk` shows that this unstratified jackknife (JK1) has 10 replicates. The jackknife standard errors for the estimated means of *tuitionfee_ in* and *tuitionfee_ out* are 454.46 and 2325.15, respectively. These are the same as the linearization standard errors because $\hat{V}_{JK}(\bar{y}) = s_y^2/n$ for an SRS, as shown in Section 9.3 of SDA. The jackknife standard error for the nonlinear statistic of the ratio, 0.231483, differs slightly from the linearization standard error of 0.231178. These values are extremely close—after all, the linearization variance and the jackknife variance are both consistent estimators for $V(\hat{B})$—but are not exactly the same. A $t$ distribution with $n - 1 = 9$ df is used for the confidence interval.

For an SRS such as this, the jackknife weights are calculated by setting the weight of observation $j$ to zero and assigning weight $w_i \times n/(n-1) = (500/10) * (10/9) = 55.555$ to each of the other observations. You can print the replicate weight multipliers for each observation if you want (although you typically do not need to do this when carrying out a data analysis).

```
# can look at replicate weight multipliers if desired
# note that observation being omitted for replicate has weight 0
# weight multiplier for other observations is 10/9 = 1.11111
round(dcollegerg1jk$repweights$weights,digits=4)
##          [,1]   [,2]   [,3]   [,4]   [,5]   [,6]   [,7]   [,8]   [,9]  [,10]
##  [1,] 0.0000 1.1111 1.1111 1.1111 1.1111 1.1111 1.1111 1.1111 1.1111 1.1111
##  [2,] 1.1111 0.0000 1.1111 1.1111 1.1111 1.1111 1.1111 1.1111 1.1111 1.1111
##  [3,] 1.1111 1.1111 0.0000 1.1111 1.1111 1.1111 1.1111 1.1111 1.1111 1.1111
##  [4,] 1.1111 1.1111 1.1111 0.0000 1.1111 1.1111 1.1111 1.1111 1.1111 1.1111
##  [5,] 1.1111 1.1111 1.1111 1.1111 0.0000 1.1111 1.1111 1.1111 1.1111 1.1111
##  [6,] 1.1111 1.1111 1.1111 1.1111 1.1111 0.0000 1.1111 1.1111 1.1111 1.1111
##  [7,] 1.1111 1.1111 1.1111 1.1111 1.1111 1.1111 0.0000 1.1111 1.1111 1.1111
##  [8,] 1.1111 1.1111 1.1111 1.1111 1.1111 1.1111 1.1111 0.0000 1.1111 1.1111
##  [9,] 1.1111 1.1111 1.1111 1.1111 1.1111 1.1111 1.1111 1.1111 0.0000 1.1111
## [10,] 1.1111 1.1111 1.1111 1.1111 1.1111 1.1111 1.1111 1.1111 1.1111 0.0000
```

**Jackknife weights for a complex survey.** For a survey with stratification and clustering, you need to include the stratum and cluster information in the *svydesign* function. The replicate weights are then constructed by deleting one psu at a time from each stratum.

**Example 9.8 of SDA.** In this example, we consider a two-stage cluster sample. Because *clutch* is the primary sampling unit (psu), the procedure deletes one clutch at a time rather than one observation at a time.

The sampling weight *relwt* is defined as $M_i/m_i = \text{csize}/2$, the number of eggs in the clutch divided by the number of eggs selected from the clutch for measurement. The *as.svrepdesign* function sets the replicate weight variable equal to 0 for all observations from the psu being deleted in that replicate. This design has one stratum, so we use **type="JK1"**; if your survey has stratification, use **type="JKn"**.

```
data(coots)
coots$relwt<-coots$csize/2
dcoots<-svydesign(id=~clutch,weights=~relwt,data=coots)
dcootsjk <- as.svrepdesign(dcoots, type="JK1")
svymean(~volume,dcootsjk)
##           mean     SE
## volume 2.4908  0.061
confint(svymean(~volume,dcootsjk),df=degf(dcootsjk))
##          2.5 %   97.5 %
## volume 2.370354 2.611203
```

The estimated mean volume is 2.49 with a standard error of 0.061 by the jackknife method. The confidence interval for the mean volume is $[2.370354, 2.611203]$, calculated using 183 (number of psus minus 1) df.

## 9.2.3 Bootstrap

**Example 9.9 of SDA.** This example looks at creating bootstrap weights using the method in Rao et al. (1992) to estimate the population distribution of height from the SRS in data set *htsrs*. The number of replicates used is specified by **replicates = 1000** in the function *as.svrepdesign* with `type = "subbootstrap"`. You will need to use *set.seed* to be able to reconstruct the same set of bootstrap weights later.

The weight variable *wt* equals 10 for each observation. In each replicate, if observation $i$ is selected $m$ times; the weight in the replicate for that observation is, using the formula in Section 9.3 of SDA, $10 \times (200/199) \times m$. The df with the bootstrap method should be set equal to the sample size minus one, which is $200 - 1 = 199$, not the number of bootstrap replicates, which can be set to any value.

```
data(htsrs)
nrow(htsrs)
## [1] 200
wt<-rep(10,nrow(htsrs))
dhtsrs<-svydesign(id=~1, weights=~wt,data=htsrs)
set.seed(9231)
dhtsrsboot <- as.svrepdesign(dhtsrs, type="subbootstrap",replicates=1000)
svymean(~height,dhtsrsboot)
##            mean      SE
## height 168.94 0.7978
degf(dhtsrsboot) # 199 = n - 1
## [1] 199
confint(svymean(~height,dhtsrsboot),df=degf(dhtsrsboot))
##            2.5 %    97.5 %
## height 167.3667 170.5133
svyquantile(~height, dhtsrsboot, quantile=c(0.25,0.5,0.75), ties=c("rounded"))
## Statistic:
##         height
## q0.25 159.70
## q0.5  168.75
## q0.75 176.00
## SE:
##           height
## q0.25 0.8878743
## q0.5  0.9930391
## q0.75 1.1339674
```

The output above shows the summary statistics of the mean, selected quantiles, and associated standard errors calculated using the bootstrap method with seed 9231.

**Example 9.10 of SDA.** The code for creating bootstrap weights in a complex survey design is similar to that for an SRS. The survey design is specified by *svydesign*, and the *as.svrepdesign* function is used to create the replicate weights. Data set *htstrat* is a stratified random sample, so *svydesign* requires only the stratum information. When the *weights* argument is not supplied, the function calculates the weights from the information in the *fpc* argument.

```
data(htstrat)
nrow(htstrat)
## [1] 200
dhtstrat <- svydesign(id = ~1, strata = ~gender, fpc = c(rep(1000,160),rep(1000,40)),
                 data = htstrat)
set.seed(982537455)
dhtstratboot <- as.svrepdesign(dhtstrat, type="subbootstrap",replicates=1000)
svymean(~height,dhtstratboot)
##            mean      SE
## height 169.02 0.7296
degf(dhtstratboot)
```

```
## [1] 198
confint(svymean(~height,dhtstratboot),df=degf(dhtstratboot))
##          2.5 %   97.5 %
## height 167.5769 170.4543
```

The estimated mean height is 169.02 with a standard error of 0.7296 by the bootstrap method. The 95% confidence interval for mean height is [167.5769, 170.4543], calculated using a $t$ distribution with 198 df (sample size minus number of strata, $200 - 2$). If you specify the bootstrap design object with a different random number seed, you will obtain a slightly different value for the standard error because a different set of bootstrap samples will be used for the variance calculations.

The *as.svrepdesign* function will also construct bootstrap replicate weights for other forms of bootstrap: `type="bootstrap"` uses the method of Canty and Davison (1999) and `type="mrbbootstrap"` uses the multistage rescaled bootstrap of Preston (2009).

### 9.2.4   Replicate Weights and Nonresponse Adjustments

The code given so far in this section constructs replicate sampling weights. When nonresponse adjustments are made to the final weights, as described in Chapter 8 of SDA, the steps of weighting class adjustments, poststratification, raking, and other adjustments that are used on the final weights need to be repeated for each replicate weight column.

For example, the combination of the *as.svrepdesign* and *postStratify* functions will create replicate weights that reflect poststratification weight adjustments. Let's look at that for the poststratified weights in Example 4.9 of SDA, which we discussed in Section 4.4.

**Example 4.9 of SDA.** Let's look at creating poststratified replicate weights for an SRS. Here are the steps:

1. Define a design object with the sampling weights, stratification, and clustering. Here, the object *dsrs* is an SRS with weights 3078/300. Because most replicate weight methods calculate the with-replacement variance, only the first-stage strata and psu information need to be supplied.

2. Choose the variance estimation method (here, JK1), and create a replicate-weights design object.

3. Apply poststratification (function *postStratify*) to the replicate-weights design object. This will poststratify the sampling weights and each variable of replicate weights.

```
data(agsrs)
# define design object for sample
dsrs <- svydesign(id = ~1, weights=rep(3078/300,300), data = agsrs)
# define replicate weights design object
dsrsjk<-as.svrepdesign(dsrs,type="JK1")
# poststratify on region
pop.region <- data.frame(region=c("NC","NE","S","W"), Freq=c(1054,220,1382,422))
dsrspjk<-postStratify(dsrsjk, ~region, pop.region)
svymean(~acres92, dsrspjk)
##            mean     SE
## acres92 299778  18653
confint(svymean(~acres92, dsrspjk),df=degf(dsrspjk))
##             2.5 % 97.5 %
## acres92 263069.2 336487
```

```
svytotal(~acres92, dsrspjk)
##              total       SE
## acres92 922717031 57413300
# Check: estimates of counts in poststrata = pop.region counts with SE = 0
svytotal(~factor(region),dsrspjk)
##                   total SE
## factor(region)NC   1054  0
## factor(region)NE    220  0
## factor(region)S    1382  0
## factor(region)W     422  0
```

The estimated mean value of *acres92* is 299,778 with a standard error of 18,653, where the jackknife replicate weights are poststratified. Recall that in example 4.9, the poststratified standard error of the mean of *acres92* is 17,513—this value is smaller because it included a finite population correction (fpc), while the replicate weight methods calculate the with-replacement variance.

Note that the final poststratified weights, and each poststratified replicate weight variable, sum to 1054 for the NC region, to 220 for the NE region, to 1382 for the S region, and to 422 for the W region—exactly the poststratification totals defined in the data set *pop.region*. After poststratification, there is no sampling variability for the variables used in the poststratification. Variables associated with the poststratification variables are expected to have reduced variance as well.

**Performing multiple steps of nonresponse adjustments.** Many surveys have several steps of weighting class adjustments followed by calibration; sometimes intermediate or final weights are trimmed or smoothed so that the weight adjustments do not have "spikes" for some observations. Each step must be repeated for each replicate sampling weight.

1. Start with the sampling weight vector $\mathbf{w}$, which is calculated as the inverse of the probability of selection for each member of the selected sample, whether respondent or nonrespondent.

2. Create $R$ replicate sampling weights $\mathbf{w}_1, \ldots, \mathbf{w}_R$ using the desired replication variance estimation method; if desired, this can be done using the *as.svyrepdesign* function.

3. Now carry out each step of the weight adjustments—weighting class adjustments, weight smoothing or trimming, propensity score adjustments, calibration, or other methods that may be used—on the sampling weight. The final weight variable results from this process.

4. Repeat the operations in Step 3 for each column of replicate weights. Each additional step in the weighting adjustments needs to be carried out separately on each replicate weight.

Some of the functions in the **survey** package, such as *postStratify*, will create replicate weights for the nonresponse adjustments they carry out. For a complicated weighting procedure, we prefer constructing the weights and replicate weights in a custom-written program so that we can see the adjustments at each step.

## 9.3    Using Replicate Weights from a Survey Data File

Section 9.2 gave examples of how to create your own replicate weights for a survey from the sampling weights, stratification, and clustering information.

Many survey organizations supply data files for which the replicate weights have already been created. These replicate weights have usually already accounted for the nonresponse weighting adjustments. You can analyze these files with the **survey** package, too.

**Example 9.5 of SDA.** Here, let's suppose that the BRR weights we created for the simple data set in Table 9.2 of SDA have been stored in an external data set and then imported into R. We create *brrdf* using the replicate weights from the design object *dbrrexbrr* that we formed in Section 9.2.1, but then use *brrdf* alone to illustrate how the function *svrepdesign* will create a design object from an imported data set that has replicate weights.

```
# Create data frame containing final and replicate weights, and y
repwts<- dbrrexbrr$repweights$weights * matrix(brrex$wt,nrow=14,ncol=8,byrow=FALSE)
brrdf<-data.frame(y=brrex$y,wt=brrex$wt,repwts)
colnames(brrdf)<-c("y","wt",paste("repwt",1:8,sep=""))
brrdf # contains weight, repwt1-repwt8, and y but no stratum info
##          y    wt repwt1 repwt2 repwt3 repwt4 repwt5 repwt6 repwt7 repwt8
## 1     2000 1500   3000      0   3000      0   3000      0   3000      0
## 2     1792 1500      0   3000      0   3000      0   3000      0   3000
## 3     4525  500   1000   1000      0      0   1000   1000      0      0
## 4     4735  500      0      0   1000   1000      0      0   1000   1000
## 5     9550  250    500      0      0    500    500      0      0    500
## 6    14060  250      0    500    500      0      0    500    500      0
## 7      800  500   1000   1000   1000   1000      0      0      0      0
## 8     1250  500      0      0      0      0   1000   1000   1000   1000
## 9     9300 1000   2000      0   2000      0      0   2000      0   2000
## 10    7264 1000      0   2000      0   2000   2000      0   2000      0
## 11   13286  250    500    500      0      0      0      0    500    500
## 12   12840  250      0      0    500    500    500    500      0      0
## 13    2106 1000   2000      0      0   2000      0   2000   2000      0
## 14    2070 1000      0   2000   2000      0   2000      0      0   2000
# create design object
dbrrdf<-svrepdesign(weights=~wt,repweights="repwt[1-9]",data=brrdf,type="BRR")
dbrrdf
## Call: svrepdesign.default(weights = ~wt, repweights = "repwt[1-9]",
##      data = brrdf, type = "BRR")
## Balanced Repeated Replicates with 8 replicates.
svymean(~y,dbrrdf) # same as before!
##       mean     SE
## y 4451.7 236.42
```

In the *svrepdesign* function, we specify the weight variable and tell the function that the replicate weights are in variables whose names are of the form *"repwt"* followed by a number. The argument `type="BRR"` indicates that the replicate weights were formed using the BRR method. The function requires only `weights=` and `repweights=` arguments to calculate point estimates and standard errors. The point estimates are calculated using the full sample weights (in the `weights=` argument), and the replicate weights are used to calculate standard errors. No stratification or clustering information is supplied to the function.

The *svrepdesign* function can also accommodate replicate weights with types JK1, JK2 (a version of jackknife for two-psu-per-stratum designs; see Brick et al., 2000), JKn, bootstrap, the Fay variant of BRR, and other structures. The producer of the survey you are analyzing will include information about the type of replicate weights that were produced and any special considerations that are needed for analysis.

## 9.4   Summary, Tips, and Warnings

Table 9.1 lists the major R functions used in this chapter.

**TABLE 9.1**
Functions used for Chapter 9.

| Function | Package | Usage |
|---|---|---|
| subset | base | Work with a subset of a vector, matrix, or data frame |
| srswor | sampling | Select an SRS without replacement |
| svydesign | survey | Specify the survey design |
| svymean | survey | Calculate mean and standard error of mean |
| svyratio | survey | Calculate ratio and standard error of ratio |
| svyquantile | survey | Calculate quantiles and their standard errors |
| as.svrepdesign | survey | Creates a replicate-weights survey design object from a design object that includes weighting, stratification, and clustering information |
| svrepdesign | survey | Creates a replicate-weights survey design object from a data frame that contains columns for final and replicate weights |

**Tips and Warnings**

- To create replicate weights for a complex survey for which stratification and clustering information is available, first create a design object with the *svydesign* function, then convert it to a replicate-weights survey design object with the *as.svrepdesign* function.

- If performing nonresponse adjustments, do the adjustments on the sampling weight variable and then on each replicated sampling weight. Then standard errors calculated using the replicate weights will account for the weighting adjustments.

- To analyze data for which replicate weights have already been supplied, create the survey design object with the *svrepdesign* function. You do not need stratification or clustering information to analyze survey data when you have the replicate weights.

- Use the *subset* function to analyze data for a domain. When using replication variance estimation, create the replicate-weights design object first, then apply the *subset* function.

# 10

# Categorical Data Analysis in Complex Surveys

The functions *svytable*, *svychisq*, and *svyloglin* from the **survey** package (Lumley, 2020) perform categorical data analyses on survey data. In this chapter, we present several examples to illustrate the usage of these functions. The code is in file `ch10.R` on the book website.

## 10.1 Contingency Tables and Odds Ratios

First, let's look at the contingency table and odds ratio for a simple random sample (SRS).

**Example 10.1 of SDA.** Data set *cablecomp* is created from the category counts in Example 10.1 of SDA. Each household in the sample gives the computer (Yes or No) and cable (Yes or No) status.

```
# create the categorical table (Table 10.1)
cablecomp<-matrix(c(119,188,88,105), ncol=2, byrow=2)
dimnames(cablecomp)<-list(Cable=c("yes", "no"),
                          Computer=c("yes","no"))
cablecomp
##         Computer
## Cable yes  no
##    yes 119 188
##    no   88 105
```

Are households with a computer more likely to subscribe to cable? A chi-square test for independence is often used for such questions. Using the function *chisq.test* (without the continuity correction), the Pearson's chi-square test statistic $X^2$ is 2.281. For large samples, $X^2$ approximately follows a chi-square ($\chi^2$) distribution with $(r-1) * (c-1)$ degrees of freedom (df) under the null hypothesis, where $r$ and $c$ are the number of rows and columns in the contingency table. In this case, $df = 1$. The $p$-value for the $X^2$ statistic is 0.13, giving no reason to doubt the null hypothesis that owning a computer and subscribing to cable television are independent.

```
# Pearson's chi-square test under multinomial sampling, obtain X^2
cablechisq<-chisq.test(cablecomp,correct=F)
cablechisq
##
##  Pearson's Chi-squared test
##
## data:  cablecomp
## X-squared = 2.281, df = 1, p-value = 0.131
# Expected values under null hypothesis
cablechisq$expected
##         Computer
## Cable      yes       no
```

```
##    yes 127.098 179.902
##    no   79.902 113.098
# odds ratio
(cablecomp[1,1]/cablecomp[1,2])/(cablecomp[2,1]/cablecomp[2,2])
## [1] 0.7552587
```

We estimate the odds of owning a computer if the household subscribes to cable as 119/188 and estimate the odds of owning a computer if the household does not subscribe to cable as 88/105. The odds ratio is therefore estimated as

$$(119/188)/(88/105) = 0.755.$$

**Contingency tables for data from a complex survey.** The only differences between constructing contingency tables and computing odds ratios for an SRS and doing so for a complex sample are that for the complex sample, we include the design information in function *svydesign* and use *svytable* to calculate the weighted counts.

**Example 10.5 of SDA.** This example shows how to use functions *svydesign* and *svytable* to produce statistics for a two-factor contingency table when observations are from a stratified multistage sample—in this case, from the Survey of Youth in Custody (*syc*) data.

The following two variables are of interest. Variable *famtime* denotes "Was anyone in your family ever incarcerated?" with 2 corresponding to No and 1 corresponding to Yes; and variable *everviol* refers to the question "Have you ever been put on probation or sent to a correctional institution for a violent offense?" with 0 corresponding to No and 1 corresponding to Yes. Next, we specify the survey design using *svydesign*, and use *svytable* to create the contingency table. Finally, we obtain a Wald chi-square test statistic with the *summary* or *svychisq* function.

```
dsyc<-svydesign(ids=~psu,weights=~finalwt,strata=~stratum,nest=TRUE,data=syc)
dsyc # Verify this is a stratified cluster sample
## Stratified 1 - level Cluster Sampling design (with replacement)
## With (861) clusters.
## svydesign(ids = ~psu, weights = ~finalwt, strata = ~stratum,
##     nest = TRUE, data = syc)
# Create contingency table by incorporating weights
example1005 <- svytable(~famtime+everviol,design=dsyc)
example1005
##         everviol
## famtime    0    1
##       1 4838 7946
##       2 4761 7154
# Perform the Wald chi-square test
summary(example1005,statistic="Wald")
##         everviol
## famtime    0    1
##       1 4838 7946
##       2 4761 7154
##
##  Design-based Wald test of association
##
## data:  svychisq(~famtime + everviol, design = dsyc, statistic = "Wald")
## F = 0.99514, ndf = 1, ddf = 845, p-value = 0.3188
# Alternatively, can calculate the Wald statistic directly using svychisq
# without forming the table first
```

```
svychisq(~famtime+everviol,design=dsyc,statistic="Wald")
##
##  Design-based Wald test of association
##
## data:  svychisq(~famtime + everviol, design = dsyc, statistic = "Wald")
## F = 0.99514, ndf = 1, ddf = 845, p-value = 0.3188
```

The Wald chi-square test statistic is $X_W^2 = 0.995$ with $p$-value of 0.32, which indicates that there is no evidence of an association between the two factors *famtime* and *everviol*.

## 10.2 Chi-Square Tests

The *svychisq* function performs all of the chi-square tests discussed in Chapter 10 of SDA. The general form of the function is:

```
svychisq(~variable1 + variable2, design.object, statistic="")
```

Table 10.1 lists some of the test statistics and measures of association produced.

**TABLE 10.1**
Chi-square test statistics calculated by function *svychisq*.

| statistic= | Statistic or Test |
|---|---|
| Wald | Wald test (Koch et al., 1975). |
| Chisq | First-order Rao–Scott test, based on Pearson's chi-square test statistic (Rao and Scott, 1981, 1984). |
| F | Second-order Rao–Scott test. |
| adjWald | Adjusted Wald test (Thomas and Rao, 1987). |
| lincom | Use exact asymptotic distribution for the linear combination of chi-square variables in the Rao–Scott statistic. |
| saddlepoint | Use saddlepoint approximation for the linear combination of chi-square variables in the Rao–Scott statistic. |

In general, we do not recommend using a finite population correction (fpc) when conducting a chi-square test. Often, the purpose of the test is to explore whether there is a general association between the factors in the superpopulation, not merely in the finite population from which the data are drawn. Conducting the test without the fpc allows generalization to the superpopulation (under some superpopulation models) while still accounting for the clustering, stratification, and unequal weights in the sampling design.

**Example 10.6 of SDA.** For this example, we define the variable *currviol* as 1 if *crimtype*=1 and 0 otherwise. This means that the "0" category of *currviol* consists of the persons with *crimtype* $\in \{2, 3, 4, 5\}$ as well as the 12 persons with missing values for *crimtype*, and can be thought of as the persons not known to have committed a violent offense. The analysis results are almost the same when the 12 missing values are excluded. We also define the variable *ageclass* as 1 if *age* is less than or equal to 15, as 2 if *age* is equal to 16 or 17, and as 3 if *age* is greater than or equal to 18.

```
# Create variables currviol and ageclass for 10.6
syc$currviol <- syc$crimtype
```

```
syc$currviol[syc$crimtype != 1 | is.na(syc$crimtype)] <- 0

syc$ageclass <- syc$age
syc$ageclass[syc$age <= 15] <- 1
syc$ageclass[syc$age == 16 | syc$age == 17] <- 2
syc$ageclass[18 <= syc$age] <- 3


# Specify the survey design
dsyc<-svydesign(ids=~psu,weights=~finalwt,strata=~stratum,nest=TRUE,data=syc)
# estimate the contingency table
svytable(~currviol+ageclass,design=dsyc)
##          ageclass
## currviol    1    2    3
##        0 4247 6542 3190
##        1 2770 4630 3633
# First-order Rao-Scott test
svychisq(~currviol+ageclass,design=dsyc,statistic="Chisq")
##
##   Pearson's X^2: Rao & Scott adjustment
##
## data:  svychisq(~currviol + ageclass, design = dsyc, statistic = "Chisq")
## X-squared = 33.993, df = 2, p-value = 0.001909
# Second-order Rao-Scott test
# (this is the default, can also request with "statistic=F")
svychisq(~currviol+ageclass, design=dsyc)
##
##   Pearson's X^2: Rao & Scott adjustment
##
## data:  svychisq(~currviol + ageclass, design = dsyc)
## F = 6.2614, ndf = 1.7258, ddf = 1458.2973, p-value = 0.003245
```

Pearson's $X^2$ statistic is 33.993. The Rao–Scott first-order test statistic adjusts $X^2$ by the design correction $\hat{E}[X^2]/2$ and is compared to a $\chi^2$ distribution with $(3-1)*(2-1) = 2$ df. The $p$-value from the Rao–Scott first-order test is 0.002.

The Rao–Scott $F$ statistic for the second-order correction is $F = 6.2614$, which is obtained by dividing the Pearson chi-square statistic ($X^2 = 33.99$) by the design correction and the adjusted numerator df. The second-order Rao–Scott test is the default for *svychisq*, or you can add `statistic = "F"` to obtain it. You may also use an exact distribution to calculate $p$-values by setting `statistic = "lincom"`; this uses a linear combination of chi-square distributions.

We do not need to calculate the design effects (deffs) for the table cells and margins in order to conduct a chi-square test because the *svychisq* function automatically adjusts for the design effects when calculating the statistics. If you would like to see the deffs, however, they can be calculated with the *svymean* function. Here, we obtain estimates of the deffs compared with simple random sampling with replacement (`deff="replace"`). The design effects for the table cells and margins are large, indicating that persons within the same psu tend to be relatively homogeneous.

```
# deffs for table cells
svymean(~interaction(factor(ageclass), factor(currviol)),design=dsyc,deff="replace")
##                                                         mean       SE    DEff
## interaction(factor(ageclass), factor(currviol))1.0  0.169798 0.028312 14.8978
## interaction(factor(ageclass), factor(currviol))2.0  0.261554 0.017127  3.9791
```

```
## interaction(factor(ageclass), factor(currviol))3.0 0.127539 0.012152  3.4771
## interaction(factor(ageclass), factor(currviol))1.1 0.110747 0.013269  4.6840
## interaction(factor(ageclass), factor(currviol))2.1 0.185111 0.019301  6.4706
## interaction(factor(ageclass), factor(currviol))3.1 0.145250 0.013478  3.8335
# deffs for table margins
svymean(~factor(ageclass)+ factor(currviol),design=dsyc,deff="replace")
##                         mean      SE     DEff
## factor(ageclass)1 0.280545 0.033395 14.4762
## factor(ageclass)2 0.446666 0.026528  7.4598
## factor(ageclass)3 0.272789 0.022366  6.6068
## factor(currviol)0 0.558892 0.025337  6.8223
## factor(currviol)1 0.441108 0.025337  6.8223
```

## 10.3   Loglinear Models

This section will discuss how to fit a loglinear model with categorical data using function *svyloglin*.

**Example 10.8 of SDA**. Recall the computer and cable data from Example 10.1 of SDA. To analyze this in the **survey** package, let's first create data set *cabledf* with 500 records—one record per observation with a sampling weight of 1, and create the survey design object.

```
cabletable <-  matrix(c(
 "no","no",105,
 "no","yes",88,
 "yes", "no",188,
 "yes", "yes",119),byrow=T,nrow=4,ncol=3)
colnames(cabletable) <- c("cable","computer","count")
cabletable <- data.frame(cabletable)
cabledf <- cabletable[rep(row.names(cabletable), cabletable[,3]), 1:2]
dim(cabledf)
## [1] 500   2
cabledf$wt <- rep(1,500)
head(cabledf)
##      cable computer wt
## 1       no       no 1
## 1.1     no       no 1
## 1.2     no       no 1
## 1.3     no       no 1
## 1.4     no       no 1
## 1.5     no       no 1
dcable <- svydesign(id=~1, weights=~wt, data=cabledf)
```

Now let's do a chi-square test for independence with the *svychisq* function. Since this is an SRS with weights of 1, *chisq.test* and *svychisq* both give the same result. The value of the chi-square statistic from *svychisq* is 2.281 with $p$-value $= 0.1314$, just as was found in the chi-square test from Example 10.1.

```
# chi-squared test for independent data, no continuity correction
chisq.test(cabledf$computer,cabledf$cable,correct=F)
##
##   Pearson's Chi-squared test
```

```
##
## data:  cabledf$computer and cabledf$cable
## X-squared = 2.281, df = 1, p-value = 0.131
svychisq(~computer+cable,design=dcable,statistic="Chisq")
##
##   Pearson's X^2: Rao & Scott adjustment
##
## data:  svychisq(~computer + cable, design = dcable, statistic = "Chisq")
## X-squared = 2.281, df = 1, p-value = 0.1314
```

We can also use the *svyloglin* function to fit an additive loglinear model with variables *cable* and *computer*.

```
# Fit loglinear model for independence, with additive factors
cableindep <- svyloglin(~factor(cable)+factor(computer),design=dcable)
summary(cableindep)
## Loglinear model: svyloglin(~factor(cable) + factor(computer), design = dcable)
##                         coef         se           p
## factor(cable)1    -0.2320788 0.04597713 4.471605e-07
## factor(computer)1  0.1737269 0.04544339 1.318753e-04
# obtain coefficients including intercept, deviance
cableindep$model
##
## Call:  glm(formula = ff, family = quasipoisson, data = dat)
##
## Coefficients:
##    (Intercept)    factor(cable)1  factor(computer)1
##         4.7866          -0.2321             0.1737
##
## Degrees of Freedom: 3 Total (i.e. Null);  1 Residual
## Null Deviance:     43.36
## Residual Deviance: 2.275  AIC: NA
# obtain the predicted counts under the independence model
cableindep$model$fitted.values
##       1       2       3       4
## 113.098 179.902  79.902 127.098
# obtain the fitted probabilities under the model
cableindep$model$fitted.values/500
##        1        2        3        4
## 0.226196 0.359804 0.159804 0.254196
```

The estimates from *summary(cableindep)* are the coefficients for *cable* = "no" and *computer* = "no". The output prints `factor(cable)1`, where 1 means "no" according to the alphabetical order of "no" and "yes". Note that the intercept here is equal to 4.786607, which is the intercept for the expected count. To calculate the fitted probabilities using the loglinear model, we need to convert it to the intercept for probabilities, which is equal to $4.786607 - \ln(500) = -1.428$. We can calculate the fitted probabilities using the formula, or by requesting the *fitted.values*, which give the predicted counts under the additive loglinear model, from the function output.

The *deviance* component of *cableindep$model* contains the deviance of 2.275. This is the value of $G^2$ from the likelihood ratio test for independence, corresponding to a $p$-value of 0.132. Alternatively, you can obtain this $p$-value from the interaction term of a saturated model or by comparing the two nested models.

```
# Fit saturated loglinear model
cablesat <- svyloglin(~factor(cable)*factor(computer),design=dcable)
summary(cablesat)
## Loglinear model: svyloglin(~factor(cable) * factor(computer), design = dcable)
##                                  coef         se           p
## factor(cable)1                -0.22106707 0.04655596 2.050156e-06
## factor(computer)1              0.15848550 0.04655596 6.635966e-04
## factor(cable)1:factor(computer)1 -0.07017373 0.04655596 1.317341e-01
# this can also be obtained by comparing the two models
anova(cablesat,cableindep)
## Analysis of Deviance Table
##  Model 1: y ~ factor(cable) + factor(computer)
## Model 2: y ~ factor(cable) + factor(computer) + factor(cable):factor(computer)
## Deviance= 2.274961 p= 0.1324977
## Score= 2.281035 p= 0.1319832
```

**Example 10.9 of SDA.** In this example, we use *svyloglin* to analyze data set *syc*. Let's first take a look at the three-way table of weighted counts for the variables *ageclass*, *everviol*, and *famtime*. We use the design object *dsyc*, defined earlier for Example 10.5 of SDA, to incorporate the weights.

```
syctable3way <- svytable(~ageclass+everviol+famtime,design=dsyc)
syctable3way
## , , famtime = 1
##
##          everviol
## ageclass    0    1
##        1 1628 2115
##        2 2332 3395
##        3  878 2436
##
## , , famtime = 2
##
##          everviol
## ageclass    0    1
##        1 1453 1723
##        2 2234 3056
##        3 1074 2375
# Estimate probabilities in table
syctable3way/sum(syctable3way)
## , , famtime = 1
##
##          everviol
## ageclass          0          1
##        1 0.06591360 0.08563100
##        2 0.09441678 0.13745496
##        3 0.03554800 0.09862747
##
## , , famtime = 2
##
##          everviol
## ageclass          0          1
##        1 0.05882829 0.06975991
##        2 0.09044901 0.12372971
##        3 0.04348354 0.09615774
```

Next, we fit a saturated model. This produces the parameter estimates, their standard errors, and the $p$-values for the null hypotheses that the individual parameters equal 0. This survey has large design effects because the facilities have a high degree of clustering with respect to ages served and severity of offenses.

The following code requests the saturated model.

```
svyloglin(~ageclass+everviol+famtime+ageclass*everviol
   + ageclass*famtime+everviol*famtime+ageclass*everviol*famtime,design=dsyc)
```

Alternatively, you could type

```
svyloglin(~ageclass*everviol*famtime,design=dsyc)
```

to include all the factors and interactions among the factors.

```
# Fit saturated loglinear model
sycsat <- svyloglin(~ageclass*everviol*famtime,design=dsyc)
summary(sycsat)
## Loglinear model: svyloglin(~ageclass * everviol * famtime, design = dsyc)
##                                   coef           se            p
## ageclass1                  -0.114863155 0.11589240 3.216275e-01
## ageclass2                   0.344093390 0.07257131 2.121916e-06
## everviol1                  -0.244591938 0.04529927 6.683913e-08
## famtime1                    0.024225036 0.03268196 4.585506e-01
## ageclass1:everviol1         0.136557159 0.05820665 1.897229e-02
## ageclass2:everviol1         0.072369272 0.03217396 2.449266e-02
## ageclass1:famtime1          0.055452399 0.03677297 1.315632e-01
## ageclass2:famtime1          0.012807329 0.02838298 6.518218e-01
## everviol1:famtime1         -0.031699074 0.02297100 1.675988e-01
## ageclass1:everviol1:famtime1 0.008882581 0.02890809 7.586381e-01
## ageclass2:everviol1:famtime1 0.016132994 0.02780612 5.617826e-01
```

The output shows the parameter estimates for the 11 terms in the saturated model, along with standard errors and $p$-values for the Wald tests that the individual parameters equal zero. There are two parameters for each term involving the three-category factor *ageclass*. The $p$-values can be used to test individual terms.

You can also compare sets of nested models. Let's fit the independent-factor model, and compare it with the saturated model.

```
# Fit additive loglinear model for independent factors
sycind <- svyloglin(~ageclass+everviol+famtime,design=dsyc)
summary(sycind)
## Loglinear model: svyloglin(~ageclass + everviol + famtime, design = dsyc)
##                  coef          se           p
## ageclass1 -0.14745454 0.11162655 1.865137e-01
## ageclass2  0.31771375 0.07232164 1.117632e-05
## everviol1 -0.22651791 0.04817065 2.571154e-06
## famtime1   0.03519814 0.03202768 2.717719e-01
# compare independent and saturated models
anova(sycsat, sycind)
## Analysis of Deviance Table
##   Model 1: y ~ ageclass + everviol + famtime
## Model 2: y ~ ageclass + everviol + famtime + ageclass:everviol + ageclass:famtime +
##      everviol:famtime + ageclass:everviol:famtime
## Deviance= 50.82273 p= 0.0007817253
## Score= 48.67106 p= 0.001132184
```

From the output, the *p*-value is 0.00113, indicating that the independence model exhibits lack of fit when compared with the saturated model.

## 10.4  Summary, Tips, and Warnings

Table 10.2 lists the major functions used in this chapter to perform categorical data analyses.

**TABLE 10.2**

Functions used for Chapter 10.

| Function | Package | Usage |
|----------|---------|-------|
| summary | base | Summarize the results from fitting a model (here used for loglinear models) |
| chisq.test | stats | Perform a chi-square test (not using survey methods) |
| anova | stats | Compute an analysis of variance table from a model object |
| predict | stats | Obtain predicted values from a model object |
| svydesign | survey | Specify the survey design |
| svymean | survey | Calculate mean and standard error of mean; also calculate design effects |
| svytable | survey | Estimate the population contingency table from survey data |
| svychisq | survey | Calculate chi-square test statistics and *p*-values for survey data |
| svyloglin | survey | Fit a loglinear model to survey data; may also be used to compare nested models |

**Tips and Warnings**

- Always look at the estimated contingency table before conducting a chi-square test or fitting a loglinear model.

- In general, we recommend conducting chi-square tests and fitting loglinear models without an fpc. It is often desired to explore the association between variables in a context more general than the particular finite population.

- Check for empty or sparse cells by applying the *table* function before analyzing the data, particularly if you are fitting a loglinear model with many terms. See Fienberg and Rinaldo (2007) for a discussion of what can go wrong when fitting loglinear models.

# 11

# Regression with Complex Survey Data

We have already used the *svyglm* function from the **survey** package (Lumley, 2020) in Chapter 4 to perform ratio and regression estimation. In this chapter, we use the function to calculate regression coefficients and provide other summary statistics for regression analyses with complex survey data. The code is in file **ch11.R** on the book website.

## 11.1 Straight Line Regression with a Simple Random Sample

For many analyses carried out on a simple random sample (SRS), results from a model-based analysis in a function designed for independent and identically distributed data (such as the *mean* function) are the same as the results from the corresponding survey analysis function (such as the *svymean* function) used with weights set equal to 1. For regression, however, the standard errors for an SRS calculated using functions *lm* or *glm*, which perform model-based regression analyses (see Section 4.6), differ from those calculated by the *svyglm* function from the **survey** package. This is because, as explained in Section 11.2 of SDA, the standard errors for the SRS calculated using linearization account for the errors in estimating the population totals of both the $x$ and $y$ variables; the model-based standard error calculated in the *glm* or *lm* function is conditional on the values of $x$ in the sample and is calculated under the model assumptions.

We can see a slight difference for the estimates calculated in Examples 11.2 and 11.4 of SDA using the *lm* and *svyglm* functions. For this example, the difference is small because the model fits well and the sample is an SRS; for other surveys and models, the difference can be greater and we recommend using the *svyglm* function to fit the model.

**Example 11.2 of SDA.** The following shows an analysis conducted using the function *lm* for the data in *anthsrs*. This conducts a model-based analysis under assumptions (A1)–(A4) given in Section 11.1 of SDA. The *summary* function shows the regression parameter estimates, standard errors, $t$ statistic and $p$-value for the null hypothesis that each parameter equals 0, value of $R^2$, and more. The function *glm*, which fits generalized linear models for non-survey data, will give the same results.

```
data(anthsrs)
fit<-lm(anthsrs$height~anthsrs$finger)
summary(fit)
##
## Call:
## lm(formula = anthsrs$height ~ anthsrs$finger)
##
## Residuals:
##     Min      1Q  Median      3Q     Max
## -3.9045 -1.1638  0.0543  1.1407  5.0543
```

DOI: 10.1201/9781003228196-11

```
##
## Coefficients:
##                 Estimate Std. Error t value Pr(>|t|)
## (Intercept)      30.3162     2.5668   11.81   <2e-16 ***
## anthsrs$finger    3.0453     0.2217   13.73   <2e-16 ***
## ---
## Signif. codes:  0 '***' 0.001 '**' 0.01 '*' 0.05 '.' 0.1 ' ' 1
##
## Residual standard error: 1.75 on 198 degrees of freedom
## Multiple R-squared:  0.4879,Adjusted R-squared:  0.4853
## F-statistic: 188.6 on 1 and 198 DF,  p-value: < 2.2e-16
```

**Example 11.4 of SDA.** Now let's perform the regression analysis using the *svyglm* function. The major difference is to include the design object created by *svydesign*, where the variable *wt* is set equal to 3000/200 for each observation.

```
anthsrs$wt<-rep(3000/200,200)
danthsrs<- svydesign(id = ~1, weight = ~wt, data = anthsrs)
degf(danthsrs)  # here, 199
## [1] 199
fit2 <- svyglm(height~finger, design=danthsrs)
summary(fit2)
##
## Call:
## svyglm(formula = height ~ finger, design = danthsrs)
##
## Survey design:
## svydesign(id = ~1, weight = ~wt, data = anthsrs)
##
## Coefficients:
##             Estimate Std. Error t value Pr(>|t|)
## (Intercept)  30.3162     2.5436   11.92   <2e-16 ***
## finger        3.0453     0.2201   13.84   <2e-16 ***
## ---
## Signif. codes:  0 '***' 0.001 '**' 0.01 '*' 0.05 '.' 0.1 ' ' 1
##
## (Dispersion parameter for gaussian family taken to be 3.046384)
##
## Number of Fisher Scoring iterations: 2
confint(fit2)    # here calculated using normal distribution
##                2.5 %    97.5 %
## (Intercept) 25.330821 35.301675
## finger       2.613917  3.476583
fit2$coefficients # contains coefficients
## (Intercept)     finger
##    30.31625    3.04525
fit2$deviance     # residual sum of squares (for this SRS example)
## [1] 606.2304
```

The values of the estimated slope and intercept are the same as in the analysis with the *lm* function, but the standard errors of the regression coefficients, here calculated using linearization, are different. In this example, where the straight-line model fits the data well, the difference in the standard errors is small. In other examples, the two sets of standard errors may exhibit wider disparities.

**Degrees of freedom (df) for regression analyses.** One other difference between the two analyses is the df to be used for the confidence intervals and hypothesis tests. For the model-based analysis in Example 11.2, a $t$ distribution with $n-$ (number of model parameters) df is used (here 198). We usually set the df equal to (number of psus) $-$ (number of strata) for regression analyses with complex survey data, regardless of the number of model parameters. One exception to that guideline is when an analysis is being done on a domain that does not appear in some of the psus (although in this situation a model-based analysis might be preferred; see Section 11.5). Valliant and Rust (2010) discuss df in survey data analyses.

**Finite population corrections for regression analyses.** If desired, you can include an *fpc* argument in the *svydesign* function to calculate standard errors that incorporate a finite population correction (fpc) and hence will be slightly smaller. We suggest omitting the *fpc* argument when performing regression analyses because we often want to learn about the relationships among variables in a universal sense (including potential populations that are similar to the finite population), not just in the particular finite population that was studied. Ask yourself: If I were estimating regression relationships for data from a population census, would I want the standard errors of the coefficients to be zero (as they would be if a census were taken because there is no sampling variability)? If the answer is no, then omit the fpc.

**Example 11.6 of SDA.** Instead of using the linearization method to calculate standard errors of the regression coefficients, we can calculate jackknife weights for the survey and use those with the *svyglm* function to compute the standard errors. The jackknife weights are $3000/199 = 15.0754$ for the observations not deleted for the replicate and 0 for the observation that is deleted.

```
danthsrsjk <- as.svrepdesign(danthsrs, type="JK1")
fit3 <- svyglm(height~finger, design=danthsrsjk)
summary(fit3)
##
## Call:
## svyglm(formula = height ~ finger, design = danthsrsjk)
##
## Survey design:
## as.svrepdesign(danthsrs, type = "JK1")
##
## Coefficients:
##             Estimate Std. Error t value Pr(>|t|)
## (Intercept)  30.3162     2.5805   11.75   <2e-16 ***
## finger        3.0453     0.2233   13.64   <2e-16 ***
## ---
## Signif. codes:  0 '***' 0.001 '**' 0.01 '*' 0.05 '.' 0.1 ' ' 1
##
## (Dispersion parameter for gaussian family taken to be 606.2304)
##
## Number of Fisher Scoring iterations: 2
confint(fit3)
##                 2.5 %    97.5 %
## (Intercept) 25.258652 35.373844
## finger       2.607591  3.482909
```

The output for the jackknife is similar to that from Example 11.4, where standard errors are calculated using linearization, but the standard errors are slightly larger. This is not a matter for concern; the two methods of variance estimation are asymptotically equivalent but often give slightly different numbers for real data sets, which are of finite size.

For comparison, Table 11.1 lists the point estimates and standard errors using *lm* and *svyglm*.

**TABLE 11.1**
Comparison of standard error estimates using linearization or jackknife for the regression coefficients produced by *glm* and *svyglm* (data *anthsrs*).

|           | Estimate | SE(lm) | SE(svyglm, linearization) | SE(svyglm, JK) |
|-----------|----------|--------|---------------------------|----------------|
| Intercept | 30.3162  | 2.5668 | 2.5436                    | 2.5805         |
| Slope     | 3.0453   | 0.2217 | 0.2201                    | 0.2233         |

## 11.2  Linear Regression for Complex Survey Data

The National Health and Nutrition Examination Survey (NHANES) data (data set *nhanes* in the SDAResources package) will be used for examples in the remaining sections. This survey has unequal weights, stratification, and clustering.

**Example 11.7 of SDA.** In this example, we fit the multiple linear regression model

$$bmxbmi = ridageyr + ridageyr^2$$

to the NHANES data; *ridageyr* is the variable in the data that gives each person's age in years. This analysis fits the model to the entire range of ages; Section 11.3 will show an example of a regression model fit to the observations in the domain of adults. To enter the design information, recall that *sdmvpsu* and *sdmvstra* are the primary sampling unit (psu) and strata variables, respectively, and *wtmec2yr* is the weight variable. We create a subset design object, *dnhanescc*, for the domain that has no missing data for the response and explanatory variables in the regression.

```
data(nhanes)
nhanes$ridageyr2<-nhanes$ridageyr^2
dnhanes<-svydesign(id=~sdmvpsu, strata=~sdmvstra,nest=TRUE,
                weights=~wtmec2yr,data=nhanes)
dnhanescc <- subset(dnhanes,complete.cases(cbind(bmxbmi,ridageyr)))
fit4<-svyglm(bmxbmi~ridageyr + ridageyr2, design=dnhanescc)
summary(fit4)
##
## Call:
## svyglm(formula = bmxbmi ~ ridageyr + ridageyr2, design = dnhanescc)
##
## Survey design:
## subset(dnhanes, complete.cases(cbind(bmxbmi, ridageyr)))
##
## Coefficients:
##                Estimate Std. Error t value Pr(>|t|)
## (Intercept) 15.2981488  0.2293566   66.70  < 2e-16 ***
## ridageyr     0.6031084  0.0188377   32.02 9.43e-14 ***
## ridageyr2   -0.0057488  0.0002311  -24.88 2.38e-12 ***
## ---
## Signif. codes:  0 '***' 0.001 '**' 0.01 '*' 0.05 '.' 0.1 ' ' 1
##
## (Dispersion parameter for gaussian family taken to be 42.25111)
```

```
##
## Number of Fisher Scoring iterations: 2
# can also extract separate elements
nobs(fit4) # number of observations used in regression
## [1] 8756
fit4$coefficients # extract regression parameters
## (Intercept)      ridageyr     ridageyr2
## 15.298148792   0.603108429  -0.005748801
1 - fit4$deviance/fit4$null.deviance  # R-squared
## [1] 0.2833541
# test linear hypotheses about model terms
regTermTest(fit4,~ridageyr + ridageyr2,df=15)
## Wald test for ridageyr ridageyr2
##  in svyglm(formula = bmxbmi ~ ridageyr + ridageyr2, design = dnhanescc)
## F =  746.9868  on  2  and  15  df: p= 9.5625e-16
```

The *svyglm* function tells R to fit a model with $y$ variable *bmxbmi* and $x$ variables *ridageyr* and *ridageyr2*. The estimated regression coefficients are accompanied by their standard errors. The fitted values are given by the equation

$$\hat{y} = 15.298 + 0.6031084 * ridageyr - 0.0057488 * ridageyr^2,$$

and the fitted values for the data observations are stored in *fit4$fitted.values*. Residuals can be found in *fit4$residuals*.

We can estimate the population value of $R^2$ using the estimated deviance from the fitted model and the null model (model fit with just an intercept term). For linear regression, $R^2$ is the proportion of variability about the mean that is explained by the regression model, which is one minus (deviance from the regression model)/(deviance from the null model). Here we estimate $R^2 = 0.2834$.

**Tests of linear hypotheses.** The *regTermTest* function gives the Wald $F$ statistic for testing the hypothesis that the terms given in the second argument are all zero. For a linear hypothesis $H_0 : \mathbf{L}\boldsymbol{\beta} = 0$, the Wald $F$ statistic equals

$$F = \frac{(\mathbf{L}\hat{\mathbf{B}})^T \left[ \mathbf{L}\,\hat{V}(\hat{\mathbf{B}})\,\mathbf{L}^T \right]^{-1} (\mathbf{L}\hat{\mathbf{B}})}{\text{rank}\left[ \mathbf{L}\,\hat{V}(\hat{\mathbf{B}})\,\mathbf{L}^T \right]}$$

(a generalized inverse may be used when the inverse of $\mathbf{L}\,\hat{V}(\hat{\mathbf{B}})\,\mathbf{L}^T$ does not exist), and is compared to an $F$ distribution with numerator df equal to the rank of $\mathbf{L}\,\hat{V}(\hat{\mathbf{B}})\,\mathbf{L}^T$. We use (number of psus minus number of strata) as the denominator df. Here, the Wald $F$ statistic of 746.99, with $p$-value much less than 0.0001, is for the null hypothesis that the coefficients of *ridageyr* and *ridageyr2* are both zero, and is compared to an $F_{2,15}$ distribution. The *regTermTest* function will perform a Rao–Scott test based on the estimated log-likelihood ratio if `method="LRT"` is specified.

**Regression analyses and missing data.** Many data sets, including the *nhanes* data, have missing values for the $y$ variable or one or more of the $x$ variables. In this case, only 8756 (=nobs(fit4)) of the 9971 observations in the data set had values for *bmxbmi*.

We performed the regression analysis on the subset of data with non-missing values. For this example, the same results would be obtained by running the regression model with design object *dnhanes* since the default method is to omit cases with missing data. For some analyses, however, entire psus might be missing the value of at least one of the model variables;

using the *subset* function ensures that the standard error calculations use information from all of the psus.

Another option is to fit the regression model with a replicate-weights design object. If you are using replication methods to estimate variances, you can analyze subsets of the data as long as the replicate weights have been created using the full sample. This is because the replicate weight construction already accounts for the full design structure. Replicate weight variance estimates are of the form $\sum_{r=1}^{R} c_r (\hat{\theta}_r - \hat{\theta})^2$, where $\hat{\theta}$ is the estimate calculated using the full-sample weight and $\hat{\theta}_r$ is the estimate calculated using the $r$th replicate weight. When you calculate domain estimates, $\hat{\theta}$ and the replicate values $\hat{\theta}_r$ are calculated using the same subset of observations, so the domain of non-missing values does not need to be defined separately.

Here we fit the model with balanced repeated replication (BRR) weights, excluding the observations with missing data. The standard errors differ slightly from those calculated earlier using linearization.

```
dnhanesbrr <- as.svrepdesign(dnhanes, type="BRR")
fit5<-svyglm(bmxbmi~ridageyr + ridageyr2, design=dnhanesbrr,na.action=na.omit)
summary(fit5)
##
## Call:
## svyglm(formula = bmxbmi ~ ridageyr + ridageyr2, design = dnhanesbrr,
##     na.action = na.omit)
##
## Survey design:
## as.svrepdesign(dnhanes, type = "BRR")
##
## Coefficients:
##              Estimate Std. Error t value Pr(>|t|)
## (Intercept) 15.2981488  0.2337378   65.45  < 2e-16 ***
## ridageyr     0.6031084  0.0190054   31.73 1.06e-13 ***
## ridageyr2   -0.0057488  0.0002321  -24.76 2.53e-12 ***
## ---
## Signif. codes:  0 '***' 0.001 '**' 0.01 '*' 0.05 '.' 0.1 ' ' 1
##
## (Dispersion parameter for gaussian family taken to be 405852.8)
##
## Number of Fisher Scoring iterations: 2
```

But be aware that item nonresponse can distort estimates of regression relationships in the population. It is also possible that when a model has many explanatory variables, each with some item nonresponse, the missing data patterns can mesh in such a way that the model is fit on relatively few observations. Or the model might be fit on data from only a few of the psus. We recommend exploring the amount and patterns of missing data before performing analyses.

**Graphing the regression equation.** As we saw in Section 7.5, the *svyplot* function will produce scatterplots that account for the survey weights. Figure 11.1 displays a bubble plot with the fitted quadratic model.

```
# plot data bmxbmi~ridageyr
svyplot(bmxbmi~ridageyr, design=dnhanes, style="bubble",basecol="gray",
        inches=0.03,xlab="Age (years)",ylab="Body Mass Index",
        xlim=c(0,80),ylim=c(10,70))
# plot fitted quadratic regression line
```

```
timevalues <- seq(2, 80, 0.02)
length(timevalues)
## [1] 3901
predicted <- predict(fit4,data.frame(ridageyr=timevalues, ridageyr2=timevalues^2))
lines(timevalues, predicted, col = "black", lwd = 3)
```

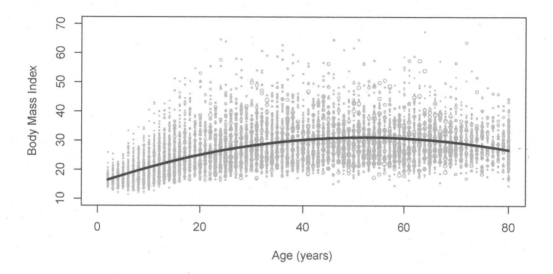

**FIGURE 11.1:** Scatterplot with fitted quadratic model.

## 11.3  Using Regression to Compare Domain Means

We computed domain means using the *svyby* function in Chapter 4. We can also compute—and compare—domain means using the *svyglm* function, by fitting a regression model with a categorical predictor variable that defines the domains.

To show how this works, let's define a few new variables and design object **dnhanes** for the remaining examples in this chapter.

- *female*: define *female* $= 1$ if *riagendr* $= 2$, and *female* $= 0$ if *riagendr* $= 1$

- *age20d*: define *age20d* $= 1$ if *ridageyr* $\geq 20$ and *bmxbmi* is not missing, and 0 otherwise

- *raceeth*: define *raceeth* = Hispanic if *ridreth3* $= 1$ or 2; *raceeth* = White if *ridreth3* $= 3$; *raceeth* = Black if *ridreth3* $= 4$; *raceeth* = Asian if *ridreth3* $= 6$; *raceeth* = Other if *ridreth3* $= 7$

- *bmi30*: define *bmi30* $= 1$ if *bmxbmi* $> 30$, *bmi30* $= 0$ if *bmxbmi* $\leq 30$

```
nhanes$female <- nhanes$riagendr-1
nhanes$age20d<-rep(0,nrow(nhanes))
nhanes$age20d[nhanes$ridageyr >=20 & !is.na(nhanes$bmxbmi)]<-1
```

```
nhanes$bmi30<-nhanes$bmxbmi
nhanes$bmi30[nhanes$bmxbmi>30]<-1
nhanes$bmi30[nhanes$bmxbmi<=30]<-0
nhanes$raceeth <- rep(NA,nrow(nhanes))
nhanes$raceeth[nhanes$ridreth3==1 | nhanes$ridreth3==2] <- "Hispanic"
nhanes$raceeth[nhanes$ridreth3==3] <- "White"
nhanes$raceeth[nhanes$ridreth3==4] <- "Black"
nhanes$raceeth[nhanes$ridreth3==6] <- "Asian"
nhanes$raceeth[nhanes$ridreth3==7] <- "Other"
# check variable construction; display missing values
table(nhanes$age20d,nhanes$bmi30,useNA="ifany")
##
##        0    1 <NA>
##   0 3130  220 1215
##   1 3248 2158    0
# no missing data for female, raceeth
table(nhanes$female,nhanes$riagendr,useNA="ifany")
##
##        1    2
##   0 4892    0
##   1    0 5079
table(nhanes$raceeth,nhanes$ridreth3,useNA="ifany")
##
##                 1    2    3    4    6    7
##   Asian         0    0    0    0 1042    0
##   Black         0    0    0 2129    0    0
##   Hispanic   1921 1308    0    0    0    0
##   Other         0    0    0    0    0  505
##   White         0    0 3066    0    0    0
dnhanes <- svydesign(id = ~sdmvpsu, strata = ~ sdmvstra, nest=TRUE,
                     weights=~wtmec2yr, data = nhanes)
```

**Example 11.8 of SDA.** Section 4.3 used the *svyby* function to calculate separate estimates for domains. The easiest way to compare domain means in the **survey** package is to fit a regression model predicting the response from one or more variables defining the domains. When there are only two domains, an indicator variable may be used. Here we fit a model on the domain of adults predicting BMI from the indicator variable *female*.

```
dnhanescc <- subset(dnhanes,!is.na(bmxbmi))
fit6<-svyglm(bmxbmi~female, design=dnhanescc)
summary(fit6)
##
## Call:
## svyglm(formula = bmxbmi ~ female, design = dnhanescc)
##
## Survey design:
## subset(dnhanes, !is.na(bmxbmi))
##
## Coefficients:
##             Estimate Std. Error t value Pr(>|t|)
## (Intercept)  26.9276     0.2070 130.076   <2e-16 ***
## female        0.6897     0.2531   2.725   0.0164 *
## ---
## Signif. codes:  0 '***' 0.001 '**' 0.01 '*' 0.05 '.' 0.1 ' ' 1
##
```

```
## (Dispersion parameter for gaussian family taken to be 58.8379)
##
## Number of Fisher Scoring iterations: 2
2*(1-pt(2.725,15)) # calculate p-value for female with 15 df
## [1] 0.01565339
```

The regression coefficient corresponding to *female* is the estimated difference between the mean BMI of females and the mean BMI of males. From the output, this value is 0.6897. The test statistic for the null hypothesis that the mean BMI is the same for both genders is $2.725 = 0.6897/0.2531$; comparing this to a $t$ distribution with 15 df (= number of psus minus number of strata) results in a $p$-value of 0.0157, indicating that the mean BMI is significantly different for the two genders. The estimated mean BMI for the male group (having *female*=0) is 26.9276, and the estimated mean BMI for the female group is 26.9276 + 0.6897 = 27.6173.

You can also see the individual domain means by fitting a regression model without an intercept with the explanatory variable *factor(female)*. The regression parameters are the estimated mean BMI for males and females.

```
fit7<-svyglm(bmxbmi~factor(female)-1, design=dnhanescc)
summary(fit7)
##
## Call:
## svyglm(formula = bmxbmi ~ factor(female) - 1, design = dnhanescc)
##
## Survey design:
## subset(dnhanes, !is.na(bmxbmi))
##
## Coefficients:
##                 Estimate Std. Error t value Pr(>|t|)
## factor(female)0   26.928      0.207   130.1   <2e-16 ***
## factor(female)1   27.617      0.260   106.2   <2e-16 ***
## ---
## Signif. codes:  0 '***' 0.001 '**' 0.01 '*' 0.05 '.' 0.1 ' ' 1
##
## (Dispersion parameter for gaussian family taken to be 58.8379)
##
## Number of Fisher Scoring iterations: 2
```

**Comparing more than two domain means.** With more than two categories, it is more convenient to declare the variable defining the domains as a categorical variable using function *factor*, and use that as the explanatory variable in the model statement of function *svyglm*. Bretz et al. (2016) discuss how to use multiple comparison methods in R to compare group means.

**Example 11.9 of SDA.** This example shows a comparison of BMI for adults in five race/ethnicity groups measured in NHANES. We defined the categories in variable *raceeth* above. We first create the subset design object to restrict the analysis to adults age 20 and over, and then fit the model with the factor variable.

```
# subset with age20d = 1 has no missing data for modeling
dnhanesadult <- subset(dnhanes,age20d==1)
fit8<-svyglm(bmxbmi~factor(raceeth), design=dnhanesadult)
summary(fit8)
##
## Call:
```

```
## svyglm(formula = bmxbmi ~ factor(raceeth), design = dnhanesadult)
##
## Survey design:
## subset(dnhanes, age20d == 1)
##
## Coefficients:
##                          Estimate Std. Error t value Pr(>|t|)
## (Intercept)               24.9707     0.1411  176.92  < 2e-16 ***
## factor(raceeth)Black       5.6205     0.3380   16.63 3.83e-09 ***
## factor(raceeth)Hispanic    5.6266     0.3368   16.71 3.65e-09 ***
## factor(raceeth)Other       5.4714     0.5337   10.25 5.76e-07 ***
## factor(raceeth)White       4.2599     0.2372   17.96 1.69e-09 ***
## ---
## Signif. codes:  0 '***' 0.001 '**' 0.01 '*' 0.05 '.' 0.1 ' ' 1
##
## (Dispersion parameter for gaussian family taken to be 47.05606)
##
## Number of Fisher Scoring iterations: 2
1 - fit8$deviance/fit8$null.deviance # R-squared
## [1] 0.03264132
# test statistic for H0: all domain means are equal
regTermTest(fit8,~factor(raceeth),df=15)
## Wald test for factor(raceeth)
##  in svyglm(formula = bmxbmi ~ factor(raceeth), design = dnhanesadult)
## F =  131.6208  on  4  and  15  df: p= 1.7277e-11
# can draw side-by-side boxplot, see Figure 11.6 of SDA for plot
# svyboxplot(bmxbmi~raceeth,dnhanesadult)
```

The above code produces estimates of the regression parameters together with standard errors and test statistics. These are the coefficients for the domain of adults ($age20d = 1$) that are shown in Table 11.5 of SDA. Note that the coefficient for the reference category, Asian, is the intercept 24.9707. The other regression coefficients estimate the difference between the mean for the category listed and the reference category, and you can calculate each estimated domain mean from those. For example, the estimated BMI for Black adults is $24.9707 + 5.620 = 30.5907$.

Alternatively, you can fit the model without an intercept. Then the estimated regression coefficients are the domain means.

```
fit9<-svyglm(bmxbmi~factor(raceeth)-1, design=dnhanesadult)
fit9$coefficients
##     factor(raceeth)Asian     factor(raceeth)Black factor(raceeth)Hispanic
##                24.97068                 30.59121                30.59725
##     factor(raceeth)Other     factor(raceeth)White
##                30.44209                 29.23055
```

Note that the estimated value of $R^2$ for this analysis is small—only about three percent of the variability is explained by the variable *raceeth*, but the Wald test shows the group means to be highly significantly different ($F = 131.6$ with $p$-value $< 0.0001$). Remember that statistical significance, which depends on the effective sample sizes for the domains, is not necessarily the same as practical importance.

## 11.4   Logistic Regression

The *svyglm* function can perform logistic regression by adding the argument `family = quasibinomial`. Call the function as:

`svyglm(y ~ x1 + x2 + ....,family=quasibinomial,design=design object)`

where $y$ is a (binary) response variables and $x1$, $x2$, ... are explanatory variables. You can specify the variance estimation methods for the *svyglm* function through the *svydesign* or *svrepdesign* functions, as discussed in Chapter 9.

**Example 11.12 of SDA.** Recall that variable *female* is defined to equal 1 if the person is a female and 0 if the person is a male. Here, *female* is treated as a numeric variable; it could also be analyzed as a categorical variable using function *factor*. Variable *bmi30* is defined as 1 if *bmxbmi* > 30, and defined as 0 if *bmxbmi* ≤ 30.

We use a subset design object to estimate parameters for adults age 20 and over who have data for all variables in the regression equation. The model fit in this example, which does not include an interaction term, has the same rate of increase for the logit of the predicted probability for males and females, even though the two genders may have different base probabilities.

```
dnhanessub2<-subset(dnhanes,age20d==1 & !is.na(bmxwaist))
lrout<-svyglm(bmi30 ~ bmxwaist+ female,family=quasibinomial,
          design=dnhanessub2)
summary(lrout)
##
## Call:
## svyglm(formula = bmi30 ~ bmxwaist + female, design = dnhanessub2,
##     family = quasibinomial)
##
## Survey design:
## subset(dnhanes, age20d == 1 & !is.na(bmxwaist))
##
## Coefficients:
##              Estimate Std. Error t value Pr(>|t|)
## (Intercept) -29.9560     1.1920 -25.130 2.09e-12 ***
## bmxwaist      0.2809     0.0115  24.434 3.00e-12 ***
## female        1.5786     0.1666   9.478 3.34e-07 ***
## ---
## Signif. codes:  0 '***' 0.001 '**' 0.01 '*' 0.05 '.' 0.1 ' ' 1
##
## (Dispersion parameter for quasibinomial family taken to be 0.7048114)
##
## Number of Fisher Scoring iterations: 7
# calculate odds ratios
exp(lrout$coefficient)
##  (Intercept)     bmxwaist        female
## 9.778316e-14 1.324332e+00 4.848338e+00
# Wald test for all parameters in model
regTermTest(lrout, ~bmxwaist+female,df=15)
## Wald test for bmxwaist female
##  in svyglm(formula = bmi30 ~ bmxwaist + female, design = dnhanessub2,
##     family = quasibinomial)
```

```
## F =  312.2534  on  2  and  15  df: p= 5.982e-13
```

The output shows the parameter estimates for the model. The odds ratio is calculated as $\exp(\hat{B}_j)$ for the corresponding regression coefficient. Thus the odds ratio for *bmxwaist* equals $\exp(0.2809) = 1.324$ and is interpreted as follows. Suppose that person 1 has a waist circumference that is 1 cm larger than the waist circumference of person 2, and the two persons have the same values for all of the other covariates in the model (in this example, that means they have the same gender). Then the model predicts the odds that person 1 has BMI > 30 to be 1.324 times as large as the odds that person 2 has BMI > 30.

The *regTermTest* function computes a Wald-type $F$ statistic for comparing nested models. Here we compute the statistic for the null hypothesis that the effects of *bmxwaist* and *female* are both 0. The Wald test statistic is $F = 312.25$ with a $p$-value much less than 0.0001, suggesting rejection of the null hypothesis.

**Graphing predicted probabilities from logistic regression.** In the following, we show a graph of the predicted probability that BMI > 30 for men and women. The predicted values from the model are stored in *lrout$fitted.values*, but we use the *predict* function to give an even distribution of points for drawing the lines.

```
waist <- seq(50,175,0.1)
xfemale <- data.frame(bmxwaist=waist,female=rep(1,length(waist)))
linfemalepred <- predict(lrout,xfemale)
xmale <- data.frame(bmxwaist=waist,female=rep(0,length(waist)))
linmalepred <- predict(lrout,xmale)
# predicted probability for female ·
femalepred <- exp(linfemalepred)/(1 + exp(linfemalepred))
# predicted probability for male
malepred <- exp(linmalepred)/(1 + exp(linmalepred))
```

```
# draw the graph
par(las=1,mar=c(4,4,1,2))
plot(waist,femalepred,type="n",xlab="Waist Circumference (cm)",
    ylab="Estimated Probability",axes=F,xlim=c(70,140))
lines(waist,femalepred,lty=1,lwd=2)
lines(waist,malepred,lty=2,lwd=2)
legend("topleft",c("Female","Male"),lty=c(1,2),bty="n")
axis(2)
axis(1)
box(bty="l")
```

The graph in Figure 11.2 shows the predicted probabilities from the model. This does not show the original data, however, and a scatterplot of a binary variable $y$ versus the explanatory variables typically does not provide much information about the relationship because it usually displays an indistinguishable mass of observations at $y = 1$ and another mass at $y = 0$. A more helpful option is to construct graphs showing the distribution of continuous covariates for each level of the response variable (here, *bmi30*). If there is a single continuous covariate, you may want to construct a histogram of the values of the covariate at each value of $y$.

Many of the diagnostics and graphs described by Allison (2012) for model-based logistic regression can also be applied to survey data.

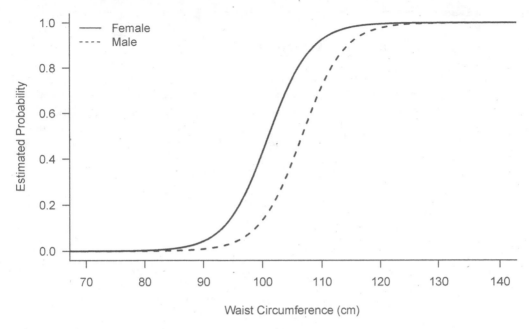

**FIGURE 11.2:** Predicted probability that BMI > 30.

## 11.5 Additional Resources and Code

**Balanced sampling: Exercise 11.37 of SDA.** There are several methods that can be used for selecting a balanced sample in R. The rejective method described in the exercise involves generating repeated samples and then selecting a sample at random from those that meet the balancing constraints within a predetermined tolerance.

The rejective method can require a great deal of computation, however. If the population and sample are large, it may not be feasible to generate repeated samples and then reject those that fail to meet the balancing criteria. The R packages `sampling` (Tillé and Matei, 2021) and `BalancedSampling` (Grafström and Lisic, 2019) use computationally efficient algorithms to select balanced samples. See also Chauvet and Tillé (2006) and Tillé and Wilhelm (2017).

**Model-based regression for survey data.** Section 11.4 of SDA discussed model-based regression analyses for survey data. You may want to see if the regression coefficients vary from cluster to cluster, or compare the results of models fit with and without weights to see if the weights have information about the regression relationships.

Packages `nlme` (Pinheiro et al., 2021) and `lme4` (Bates et al., 2015, 2020) are commonly used to fit mixed models in R. Functions from these packages can be used to fit linear and generalized mixed models to survey data, using a model-based approach in which stratification variables can be included as covariates and clusters can be included as random effects. As of this writing, these packages do not provide options for fitting mixed models from a design-based perspective.

## 11.6   Summary, Tips, and Warnings

The *svyglm* function performs linear and logistic regression analyses with numeric and categorical explanatory variables. It will also fit other generalized linear models to survey data. When called without the *family* argument or with `family=gaussian`, it will fit a linear regression model; when called with `family=quasibinomial`, it will fit a logistic regression model; when called with `family=quasipoisson`, it will fit a Poisson regression model.

Table 11.2 lists *svyglm* and the other major functions used in this chapter to perform regression analyses.

**TABLE 11.2**
Functions used for Chapter 11.

| Function | Package | Usage |
|---|---|---|
| summary | base | Summarize the results from fitting a model |
| subset | base | Work with a subset of a vector, matrix, or data frame |
| predict | stats | Obtain predicted values from a model object |
| confint | stats | Calculate confidence interval |
| lm | stats | Fit a linear model to a data set (not using survey methods) |
| anova | stats | Compute an analysis of variance table from a model object |
| nobs | stats | Find out how many observations were used to fit a model |
| svydesign | survey | Specify the survey design |
| as.svrepdesign | survey | Create a replicate-weights survey design object from a design object that includes weighting, stratification, and clustering information |
| svyboxplot | survey | Draw boxplot of survey data, incorporating the weights |
| svyplot | survey | Draw scatterplot of survey data, incorporating the weights |
| svyglm | survey | Fit a generalized linear model to survey data |
| regTermTest | survey | Calculate Wald test statistic from model object created using *svyglm* |

The main commands used for a typical analysis with the *svyglm* function are given below. Many other statements and options are available for the *svyglm* function, and these are described in the **survey** package documentation.

```
# categorical variable using function 'factor'
mydata$class_var1<-factor(mydata$class_var1)
mydata$class_var2<-factor(mydata$class_var2)
# enter design object
dobject <- svydesign(id = ~psu, strata = ~ strata, nest=TRUE,
                     weights=~wt, data = mydata)
# enter replicate methods for variance calculation if desired
# or use svrepdesign to use data already containing replicate weights
dobject2 <- as.svrepdesign(dobject, type=" ")
# enter domain info, if analysis is restricted within domain
dobjectsub<-subset(dobject, domain = )
```

```
# or using replicate methods for variance calculation
dobjectsub2<- subset(dobject2, domain = )
# perform regression analysis
# design = dobject for whole data analysis using linearization
# design = dobject2 for whole data analysis using replicate methods
# design = dobjectsub for domain analysis using linearization
# design = dobjectsub2 for domain analysis using replicate methods
fit<-svyglm(y ~x1+ x2+ class_var1 + class_var2, design= )
# logistic regression
fit<-svyglm(y ~x1+ x2+ class_var1 + class_var2,
            family=quasibinomial, design= )
# Display coefficient estimatets etc
summary(fit)
# test linear hypotheses using Wald statistic
regTermTest(fit, ~variable1 + variable2)
# estimate R-squared
1 - fit$deviance/fit$null.deviance
# extract coefficients
fit$coefficients
# extract residuals, fitted values
resid <- fit$residuals
fitted.values <- fit$fitted.values
```

For replication variance estimation, specify which variance estimation method is used in the *type* argument of function *as.svrepdesign* or *svrepdesign*, as described in Chapter 9.

## Tips and Warnings

- If your data set has item nonresponse, function *nobs* will tell you how many observations were used in the analysis. If many observations were excluded from the model because they were missing $y$ or one of the $x$ variables, you might want to consider an alternative model for the data. You may also want to investigate the pattern of missing data across psus.

- If separate regression models are desired for domains, use the *subset* function to define design objects for the domains.

- You can use the *svyby* (see Section 4.3) or *svyglm* functions to compute domain means.

# 12

## Additional Topics for Survey Data Analysis

In this chapter, we present some methods for analyzing simple two-phase samples and estimating the size of a population using capture-recapture methods. We also describe resources for fitting more complex models to estimate population size, and for fitting small area estimation models. The code is in file `ch12.R` on the book website.

## 12.1 Two-Phase Sampling

In a two-phase design, the final sample is taken in two steps. First, a sample is selected from a population using a probability sampling design. In the second step, information from the first-phase sample is used to set selection probabilities for a subsample to be drawn from the first-phase sample. The function *twophase* from the **survey** package (Lumley, 2020) will analyze data from two-phase samples.

**Examples 12.1 and 12.4 of SDA.** In this example, we use two-phase sampling to estimate the percentage of Vietnam-era veterans in U.S. Veterans Administration (VA) hospitals who actually served in Vietnam (Stockford and Page, 1984). In the first phase, the 1982 VA Annual Patient Census (APC) included a random sample of 20% of the patients in VA hospitals. Therefore, *phase1wt* is equal to 5. After the phase I data were collected, the answers to the question "Was military service in Vietnam?" ("Yes," "No," or "Not Available") were obtained from medical records and used to determine strata for the second phase of sampling. The second phase obtained the true classification of Vietnam service from the military records of a stratified random subsample of the phase I sample. Variable *phase2wt* gives the stratified weights from the second phase of sampling.

The results for the question "Was service in Vietnam?" for this survey are given in Table 12.1.

**TABLE 12.1**

Results for the question, "Was service in Vietnam?"

| APC Group | APC Classification | Phase II Sample Size | Vietnam Service |
|-----------|--------------------|-----------------------|-----------------|
| Yes       | 755                | 67                    | 49              |
| No        | 804                | 72                    | 11              |
| NotAvail  | 505                | 505                   | 211             |
| Total     | 2064               | 644                   | 271             |

From Table 12.1, the percentage of veterans with Vietnam service differed for the three groups: Of the veterans with a "yes" response to the APC survey question, 73% actually served in Vietnam, compared with 15% for the "no" group and 42% for the veterans for whom the information was not available.

DOI: 10.1201/9781003228196-12

We can use the *twophase* function to estimate the population percentage of persons who served in Vietnam. The function has the form

```
twophase(id=list(~ ,~), strata = list(~, ~) , probs = list(~, ~),
        weights = list(~, ~), fpc = list(~, ~), subset,
        data, method=c("full","approx","simple"))
```

where the *id*, *strata*, *probs*, *weights*, and *fpc* arguments (usually, only some of these arguments are needed) are supplied with information for both design phases. The *subset* argument is used to specify which observations are selected in phase 2. Three methods are available for variance estimation. The `"full"` method, which requires the sampling probabilities for each stage, gives unbiased variance estimates for general multistage designs. The `"simple"` and `"approx"` methods are simpler and use less memory: these will calculate variances for designs in which an SRS is taken at phase I and a stratified random sample is taken at phase II.

Many two-phase surveys, such as the survey considered here and the majority of the surveys in the exercises for Chapter 12 of SDA, have an SRS at phase I and a stratified sample at phase II, so we show how to use the *twophase* function with the `"simple"` method for variance estimation.

The *id*, *weights*, and *strata* arguments to *twophase* are lists:

- We enter the weights at each phase as `weights=list(~phase1wt, ~phase2wt)`. For this application, each phase I weight, in variable *phase1wt*, equals $N$ divided by the phase I sample size; the phase II weights in variable *phase2wt* are $n_h/m_h$, the observed phase I sample size in stratum $h$ divided by the phase II sample size in stratum $h$.

- We use `id=list(~1,~1)` to show that there is no clustering at either phase of sampling.

- The phase I sample is an SRS, and the phase II sample is stratified by the information in variable *apc* that is gathered at phase I. This is entered as `strata=list(NULL,~apc)`, where the NULL for phase I indicates no stratification was used at that phase of sampling.

Only the records sampled in phase II have information for *vietnam*, the variable of interest. We thus analyze only the subset of records from the phase II sample; these are the records having *vietnam$p2sample*=1.

```
data(vietnam)
print.data.frame(vietnam[1:6,])
##   apc p2sample vietnam phase1wt phase2wt  finalwt p1apcsize p2apcsize
## 1 Yes        1       1       5 11.26866 56.34328       755        67
## 2 Yes        1       1       5 11.26866 56.34328       755        67
## 3 Yes        1       1       5 11.26866 56.34328       755        67
## 4 Yes        1       1       5 11.26866 56.34328       755        67
## 5 Yes        1       1       5 11.26866 56.34328       755        67
## 6 Yes        1       1       5 11.26866 56.34328       755        67
nrow(vietnam)   #2064
## [1] 2064
# define logical variable to specify which obsns are selected in phase 2
vietnam$indexp2<- vietnam$p2sample==1
dphase2<-twophase(id=list(~1,~1), weights=list(~phase1wt, ~phase2wt),
        strata=list(NULL,~apc), subset=~indexp2, data=vietnam, method="simple")
svymean(~vietnam, dphase2)
##            mean     SE
## vietnam 0.42926 0.0271
```

From the two-phase sample, the estimated percentage of Vietnam-era veterans in U.S. Veterans Administration hospitals who actually served in Vietnam is 0.42926 with a standard error of 0.0271.

---

## 12.2 Estimating the Size of a Population

### 12.2.1 Ratio Estimation of Population Size

As discussed in Section 13.1 of SDA, the simple two-sample capture-recapture estimate can be calculated using ratio estimation. Symmetric confidence intervals calculated using the $t$ distribution can have poor coverage probability in small samples, so we also discuss calculating confidence intervals using inverted likelihood-ratio tests and bootstrap.

**Example 13.1 of SDA.** In this example, we estimate the total number of fish $N$ in a lake together with the standard error based on a capture-recapture sample.

In the first step, catch and mark 200 fish in the lake, then release them. Allow the marked and released fish to mix with the other fish in the lake. Next, take a second, independent sample of 100 fish. Suppose that 20 of the fish in the second sample are marked. Assuming that the population of fish has not changed between the two samples and that each catch gives a simple random sample (SRS) of fish in the lake. Below is the data set illustrated in a $2 \times 2$ table.

**TABLE 12.2**
Data for Example 13.1 of SDA.

|  |  | In Sample 2 | |  |
|---|---|---|---|---|
|  |  | Yes | No |  |
| **In Sample 1** | Yes | 20 | 180 | 200 |
|  | No | 80 | ? | ? |
|  |  | 100 | ? | N |

The function *svyratio* in the **survey** package can be used to estimate the population total with a symmetric confidence interval. We create a data set with records for the second sample, where every observation has a weight of 1.

```
# create data frame of records from sample 2 of size n2
n1 <- 200
n2 <- 100
m <- 20
fish<-data.frame(x=c(rep(1,m),rep(0,n2-m)),wt=rep(1,n2),n1=rep(n1,n2))
dfish<-svydesign(id=~1,weights=~wt,data=fish)
estpop<-svyratio(~n1,~x,dfish)
estpop
## Ratio estimator: svyratio.survey.design2(~n1, ~x, dfish)
## Ratios=
##        x
## n1 1000
## SEs=
##             x
## n1 201.0076
```

```
# calculate symmetric confidence interval
confint(estpop,df=n2-1)
##            2.5 %    97.5 %
## n1/x 601.1574 1398.843
```

For many applications, however, the distribution of the ratio is skewed so that a symmetric confidence interval may not have accurate coverage probability. The function *captureci* from the **SDAResources** package (Lu and Lohr, 2021) will compute a confidence interval using the method of Cormack (1992).

To apply function *captureci*, first construct a matrix *xmat*, where $1 =$ "in sample" and $0 =$ "not in sample." For our example, *xmat* has two columns since there are two samples; the row $(0,0)$ represents the category of not being in either sample. Next, we define $y$ with the number of fish corresponding to *xmat*. Table 12.3 illustrates the definition of *xmat* and $y$.

### TABLE 12.3
Matrix notation for Example 13.1 of SDA.

| Description | Vector Notation, xmat | Number of Fish, $y$ |
|---|---|---|
| In sample 1 and sample 2 | $(1,1)$ | 20 |
| In sample 1, not in sample 2 | $(1,0)$ | 180 |
| Not in sample 1, in sample 2 | $(0,1)$ | 80 |

After defining *xmat* and $y$, simply type `captureci(xmat,y)` to derive the estimates.

```
# define xmat and y
xmat <- cbind(c(1,1,0),c(1,0,1))
y <- c(20,180,80)
cbind(xmat,y) # show xmat and y
##            y
## [1,] 1 1  20
## [2,] 1 0 180
## [3,] 0 1  80
captureci(xmat,y)
## $cell
## (Intercept)
##         720
##
## $N
## (Intercept)
##        1000
##
## $CI_cell
## [1]   436.199 1233.835
##
## $CI_N
## [1]   716.199 1513.835
##
## $deviance
## [1] 1.598721e-14
```

The function *captureci* reports the estimated *cell* value for the missing count for category $(0,0)$, the estimated total $\hat{N}$, and the confidence intervals for the missing category count. The estimated total fish in the lake is 1000, with a 95% confidence interval of $[716.199, 1513.835]$.

We can also derive a confidence interval for the population size using the bootstrap method. The first column of data *fish* contains a vector of length $n_2$ that indicates membership in sample 1. We can take $R$ samples with replacement from this vector, calculate Chapman's (1951) estimate

$$\tilde{N} = \frac{(n_1 + 1)(n_2 + 1)}{m + 1} - 1$$

from each resample, and then use the $R$ estimates from the bootstrap resamples to estimate the sampling distribution of $\tilde{N}$. We use Chapman's estimate here because it is guaranteed to be finite for every bootstrap resample, whereas $\hat{N} = n_1 n_2 / m$ might be infinite if a resample contains no marked fish.

```
chapman<-function(y,n1) { (n1+1)*(length(y)+1)/(sum(y)+1) - 1}
Ntilde<-chapman(fish[,1],n1)
Ntilde
## [1] 965.7143
# generate 2000 bootstrap samples
nboot<-2000
set.seed(9231)
bootsamp<-matrix(sample(fish[,1], size = nrow(fish)*nboot, replace=TRUE), ncol=nboot)
dim(bootsamp) # nboot columns of resamples
## [1]  100 2000
# calculate Chapman's estimate for each column
Ntildeboot<-apply(bootsamp,2,chapman,n1)
# draw histogram of bootstrap distribution
par(las=1)
hist(Ntildeboot,xlab = "Estimated Population Size",
  main = "Histogram of Bootstrap Estimates",col="lightgray",
  breaks=20,border="white")
box(bty="l")
# calculate percentiles to get confidence interval
quantile(Ntildeboot,probs=c(0.025,0.975))
##      2.5%     97.5%
##  699.0345 1560.6154
```

The estimated confidence interval for total fish in the lake, using the bootstrap method with a seed of 9231, is [699, 1561]. The histogram in Figure 12.1 shows the distribution of the estimates from the bootstrap replicates. Note that the distribution is skewed, indicating that a symmetric confidence interval, produced under the assumption that $\tilde{N}$ follows a $t$ distribution, is not a good choice for this data set.

## 12.2.2 Loglinear Models with Multiple Lists

Section 10.3 used the *svyglm* function to fit loglinear models to data from a complex survey. When loglinear models are used in multiple-recapture estimation, it is often assumed that the lists are simple random samples. This section uses the function *captureci* from package **SDAResources** to calculate confidence intervals for the missing cell and the population size. In *captureci*, the loglinear model is fit using Poisson regression with the *glm* function.

**Example 13.3 of SDA.** Domingo-Salvany et al. (1995) used capture–recapture to estimate the prevalence of opiate addiction in Barcelona, Spain. One of their data sets consisted of three samples from 1989: (1) a list of opiate addicts from emergency rooms (E list), (2) a list of persons who started treatment for opiate addiction during 1989, reported to the Catalonia Information System on Drug Abuse (T list), and (3) a list of heroin overdose

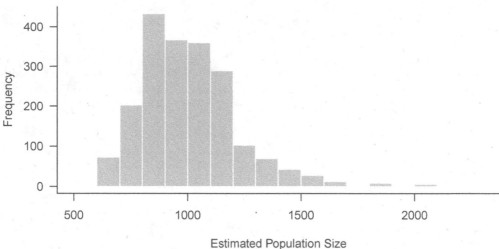

**FIGURE 12.1:** Histogram of population size estimates from bootstrap replicates.

deaths registered by the forensic institute in 1989 (D list). A total of 2864 distinct persons were on the three lists.

Persons on the three lists were matched, with the results listed in Table 12.4. The Membership column of the table gives the group membership in binary notation, with 1 denoting membership in the list. There are 712 observations on the T list but neither of the others, 69 observations in the D list but neither of the others, 314 observations on both the E list and the T list but not on the D list, and so on, with 6 observations on all three lists. The first line of the table shows a missing count for the units not on any of the lists.

**TABLE 12.4**
Notation for Example 13.3 of SDA.

| List | Membership (E List, D List, T List) | Count |
|------|-------------------------------------|-------|
| None | (0,0,0) | ? |
| T | (0,0,1) | 712 |
| D | (0,1,0) | 69 |
| E | (1,0,0) | 1728 |
| DT | (0,1,1) | 8 |
| ET | (1,0,1) | 314 |
| ED | (1,1,0) | 27 |
| EDT | (1,1,1) | 6 |

Using Table 12.4, we can define `xmat2` and $y2$, and apply function *captureci* to derive the estimates.

```
# define xmat2 and y2
xmat2<-cbind(c(1,1,1,0,1,0,0),c(1,1,0,1,0,1,0),c(1,0,1,1,0,0,1))
y2 <- c(6,27,314,8,1728,69,712)
# apply captureci
```

```
captur=ci(xmat2,y2)
## $cell
## (Intercept)
##     3966.743
##
## $N
## (Intercept)
##     6830.743
##
## $CI_cell
## [1] 3461.950 4547.747
##
## $CI_N
## [1] 6325.950 7411.747
##
## $deviance
## [1] 1.797782
```

From the output, the number of persons who have opiate addiction in Barcelona, Spain is estimated as 6830.743, with a CI of $[6325.95, 7411.747]$.

Several contributed packages in R will fit these, and more complicated, capture-recapture models. The `Rcapture` package (Baillargeon and Rivest, 2007; Rivest and Baillargeon, 2019) estimates population sizes for open and closed populations using loglinear models, and will compute all of the estimates discussed in SDA (and more). The `SpadeR` (Species Prediction and Diversity Estimation) package (Chao et al., 2016) fits a variety of models for estimating population sizes and biodiversity indices.

---

## 12.3   Small Area Estimation

Many researchers have implemented algorithms for computing small area estimates (SAEs). Pratesi (2016) and Tzavidis et al. (2018) describe some of the macros and packages that have been developed to fit small area models with SAS and R software. The basic area-level and unit-level models described in SDA can be fit using the `nlme` (Pinheiro et al., 2021) or `lme4` (Bates et al., 2015, 2020) packages in R.

Numerous contributed packages are available for computing small area estimates in R; Rao and Molina (2015), Hidiroglou et al. (2019), and Kreutzmann et al. (2019) describe some of the packages that have been developed. These packages calculate SAEs for a variety of situations—for example, accounting for spatial or temporal correlation among areas, or for measurement error among the covariates; using robust estimation of model parameters; or adopting a fully Bayesian approach.

The following list gives brief descriptions of three packages that have been used to produce small area estimates in various applications. All also compute the mean squared errors of the estimates.

- Package `sae`, "Small Area Estimation" (Molina and Marhuenda, 2015, 2020), calculates SAEs using a variety of models, including the Fay-Herriot model and the basic unit-level model. Models are also available that incorporate spatial or temporal information.

- Package `emdi`, "Estimating and Mapping Disaggregated Indicators" (Kreutzmann et al., 2019; Harmening et al., 2021), finds SAEs for small geographic areas, with an emphasis on poverty indicators; the package also includes a variety of diagnostic plots.

- Package `mme`, "Multinomial Mixed Effects Models" (Lopez-Vizcaino et al., 2019), fits Gaussian-multinomial models to calculate SAEs of proportions, accounting for temporal correlations.

The Asian Development Bank (2020) provides a step-by-step guide to calculating small area estimates in R, using the `sae` and `survey` packages, with code for computing and creating maps for small area poverty estimates.

## 12.4  Summary

Table 12.5 lists the major functions used in this chapter.

**TABLE 12.5**
Functions used for Chapter 12.

| Function | Package | Usage |
|---|---|---|
| sample | base | Select with-replacement samples for use with bootstrap |
| confint | stats | Calculate confidence interval |
| glm | stats | Fit a generalized linear model (not using survey methods) |
| svydesign | survey | Specify the survey design |
| svyratio | survey | Calculate a ratio and its standard error from a survey |
| twophase | survey | Calculate estimates and standard errors from a two-phase survey |
| captureci | SDAResources | Calculate a confidence interval for population size using an inverted likelihood-ratio test |

**Continuing the journey.** The capabilities of R continue to expand as new statistical methods are developed and implemented in packages. You can find recent contributions under "Packages" on the left panel of the website `https://cran.r-project.org/`.

Still have questions after reading this book? There are many resources available online that provide help for using R. The websites at

```
https://support.rstudio.com/hc/en-us/articles/200552336-Getting-Help-with-R
https://www.r-bloggers.com
https://journal.r-project.org/
```

provide links to articles and pages where members of the R user community post answers to questions. The chances are that you are not the first to have your question and that someone has posted an answer to it online.

# A

Data Set Descriptions

The data sets referenced in SDA and described in this appendix are available from the book website (see the Preface for the website address) and in the contributed R package SDAResources (Lu and Lohr, 2021). These data sets are provided for instructional purposes only and without warranty. Anyone wishing to investigate the subject matter further should obtain the original data from the source. In some cases, the data sets referenced in SDA and this book are a subset of the original data; in others, the information has been modified to protect the confidentiality of the respondents.

All data sets ending in .csv use commas as a separator between fields.

These data sets have also been stored in SAS format (with the name ending in .sas7bdat) and R format with missing values recoded to the symbols used for missing data in the software package ('.' or blank in SAS and NA in R).

**agpop.csv**    Data from the 1992 U.S. Census of Agriculture. Source: U.S. Bureau of the Census (1995). In columns 3–14, the value −99 denotes missing data.

| Column | Name | Value |
|--------|------|-------|
| 1 | county | county name (character variable) |
| 2 | state | state abbreviation (character variable) |
| 3 | acres92 | number of acres devoted to farms, 1992 |
| 4 | acres87 | number of acres devoted to farms, 1987 |
| 5 | acres82 | number of acres devoted to farms, 1982 |
| 6 | farms92 | number of farms, 1992 |
| 7 | farms87 | number of farms, 1987 |
| 8 | farms82 | number of farms, 1982 |
| 9 | largef92 | number of farms with 1,000 acres or more, 1992 |
| 10 | largef87 | number of farms with 1,000 acres or more, 1987 |
| 11 | largef82 | number of farms with 1,000 acres or more, 1982 |
| 12 | smallf92 | number of farms with 9 acres or fewer, 1992 |
| 13 | smallf87 | number of farms with 9 acres or fewer, 1987 |
| 14 | smallf82 | number of farms with 9 acres or fewer, 1982 |
| 15 | region | S = south; W = west; NC = north central; NE = northeast |

**agpps.csv**    Data from a without-replacement probability-proportional-to-size sample from file agpop.csv.

DOI: 10.1201/9781003228196-A

| Column | Name | Value |
|---|---|---|
| 1 | county | county name |
| 2 | state | state abbreviation |
| 3 | acres92 | number of acres devoted to farms, 1992 |
| 4 | acres87 | number of acres devoted to farms, 1987 |
| 5–15 | ... | same as variables 5–15 in `agpop.csv` |
| 16 | sizemeas | size measure used to select the pps sample |
| 17 | SelectionProb | inclusion probability for county $i$, $\pi_i$ |
| 18 | SamplingWeight | sampling weight for county $i$, $w_i = 1/\pi_i$ |
| 19 | Unit | unit number for indexing joint inclusion probabilities |
| 20–34 | JtProb_1– JtProb_15 | columns of joint inclusion probabilities |

**agsrs.csv**   Data from an SRS of size 300 from the 1992 U.S. Census of Agriculture. Variables are the same as in `agpop.csv`. In columns 3–14, the value $-99$ denotes missing data.

**agstrat.csv**   Data from a stratified random sample of size 300 from the 1992 U.S. Census of Agriculture data in `agpop.csv`. In columns 3–14, the value $-99$ denotes missing data.

| Column | Name | Value |
|---|---|---|
| 1 | county | county name |
| 2 | state | state abbreviation |
| 3 | acres92 | number of acres devoted to farms, 1992 |
| 4 | acres87 | number of acres devoted to farms, 1987 |
| 5 | acres82 | number of acres devoted to farms, 1982 |
| 6 | farms92 | number of farms, 1992 |
| 7 | farms87 | number of farms, 1987 |
| 8 | farms82 | number of farms, 1982 |
| 9 | largef92 | number of farms with 1,000 acres or more, 1992 |
| 10 | largef87 | number of farms with 1,000 acres or more, 1987 |
| 11 | largef82 | number of farms with 1,000 acres or more, 1982 |
| 12 | smallf92 | number of farms with 9 acres or fewer, 1992 |
| 13 | smallf87 | number of farms with 9 acres or fewer, 1987 |
| 14 | smallf82 | number of farms with 9 acres or fewer, 1982 |
| 15 | region | S = south; W = west; NC = north central; NE = northeast |
| 16 | rn | random numbers used to select sample in each stratum |
| 17 | strwt | sampling weight for each county in sample |

**algebra.csv**   Hypothetical data for an SRS of 12 algebra classes in a city, from a population of 187 classes.

| Column | Name | Value |
|---|---|---|
| 1 | class | Class number |
| 2 | Mi | Number of students ($M_i$) in class |
| 3 | score | Score of student on test |

**anthrop.csv**  Finger length and height for 3,000 criminals. Source: Macdonell (1901). This data set contains information for the entire population.

| Column | Name | Value |
|--------|------|-------|
| 1 | finger | length of left middle finger (cm) |
| 2 | height | height (inches) |

**anthsrs.csv**  Finger length and height for an SRS of size 200 from anthrop.csv.

| Column | Name | Value |
|--------|------|-------|
| 1 | finger | length of left middle finger (cm) |
| 2 | height | height (inches) |
| 3 | wt | sampling weight |

**anthuneq.csv**  Finger length and height for a with-replacement unequal-probability sample of size 200 from `anthrop.csv`. The probability of selection, $\psi_i$, was proportional to 24 for $y < 65$, 12 for $y = 65$, 2 for $y = 66$ or 67, and 1 for $y > 67$.

| Column | Name | Value |
|--------|------|-------|
| 1 | finger | length of left middle finger (cm) |
| 2 | height | height (inches) |
| 3 | wt | sampling weight |

**artifratio.csv**  Values from all possible SRSs for artificial population in Chapter 4 of SDA.

| Column | Name | Value |
|--------|------|-------|
| 1 | sample | sample number |
| 2 | i1 | first unit in sample |
| 3 | i2 | second unit in sample |
| 4 | i3 | third unit in sample |
| 5 | i4 | fourth unit in sample |
| 6 | xbars | $\bar{x}_{\mathcal{S}}$ |
| 7 | ybars | $\bar{y}_{\mathcal{S}}$ |
| 8 | bhat | $\hat{B}$ |
| 9 | tSRS | $\hat{t}_{y,\text{SRS}} = N\bar{y}_{\mathcal{S}}$ |
| 10 | thatr | $\hat{t}_{yr}$ |

**asafellow.csv**  Information from a stratified random sample of Fellows of the American Statistical Association elected between 2000 and 2018. The list of Fellows serving as the population was downloaded from `https://www.amstat.org/ASA/Your-Career/Awards/ASA-Fellows-list.aspx` on March 18, 2019. All other information was obtained from public sources.

| Column | Name | Value |
|--------|------|-------|
| 1 | awardyr | Year of award |
| 2 | gender | Gender of Fellow (character variable, M = male, F = female) |

| | asafellow.csv (continued) | |
|---|---|---|
| **Column** | **Name** | **Value** |
| 3 | popsize | Population size in stratum ($= N_h$) |
| 4 | sampsize | Sample size in stratum ($= n_h$) |
| 5 | field | Field of employment (character variable) acad = academia, ind = industry, govt = government |
| 6 | degreeyr | Year in which Fellow received terminal degree (year of Ph.D. if applicable, otherwise year of Master's or Bachelor's degree) |
| 7 | math | = 1 if majored in mathematics as undergraduate, 0 if did not major in math, −99 if missing |

**auditresult.csv**  Audit data used in Chapter 6 of SDA.

| **Column** | **Name** | **Value** |
|---|---|---|
| 1 | account | audit unit |
| 2 | bookvalue | book value of account |
| 3 | psi | probability of selection |
| 4 | auditvalue | audit value of account |

**auditselect.csv**  Selection of accounts for audit data used in Chapter 6 of SDA.

| **Column** | **Name** | **Value** |
|---|---|---|
| 1 | account | audit unit |
| 2 | bookval | book value of account |
| 3 | cumbv | cumulative book value |
| 4 | rn1 | random number 1 selecting account |
| 5 | rn2 | random number 2 selecting account |
| 6 | rn3 | random number 3 selecting account |

**azcounties.csv**  Population and housing unit estimates for Arizona counties (excluding Maricopa and Pima counties), from the American Community Survey 2018 5-year estimates. Source: https://data.census.gov/, accessed November 27, 2020.

| **Column** | **Name** | **Value** |
|---|---|---|
| 1 | number | County number |
| 2 | name | County name (character variable, length 15) |
| 3 | population | Population estimate for county |
| 4 | housing | Housing unit estimate for county |
| 5 | ownerocc | Number of owner-occupied housing units for county |

**baseball.csv**  Statistics on 797 baseball players, compiled by Jenifer Boshes from the rosters of all major league teams in November 2004. Source: Forman (2004). Missing values (for variables *pball, intwalk, hbp,* and *sacrfly*; all other variables have complete data) are coded as −9.

| Column | Name | Value |
|---|---|---|
| 1 | team | team played for at beginning of the season |
| 2 | leagueid | AL or NL |
| 3 | player | a unique identifier for each baseball player |
| 4 | salary | player salary in 2004 |
| 5 | pos | primary position coded as P, C, 1B, 2B, 3B, SS, RF, LF, or CF |
| 6 | gplay | games played |
| 7 | gstart | games started |
| 8 | inning | number of innings |
| 9 | putout | number of putouts |
| 10 | assist | number of assists |
| 11 | error | Errors |
| 12 | dplay | number of double plays |
| 13 | pball | number of passed balls (only applies to catchers) |
| 14 | gbat | number of games that player appeared at bat |
| 15 | atbat | number of at bats |
| 16 | run | number of runs scored |
| 17 | hit | number of hits |
| 18 | secbase | number of doubles |
| 19 | thirdbase | number of triples |
| 20 | homerun | number of home runs |
| 21 | rbi | number of runs batted in |
| 22 | stolenb | number of stolen bases |
| 23 | csteal | number of times caught stealing |
| 24 | walk | number of times walked |
| 25 | strikeout | number of strikeouts |
| 26 | intwalk | number of times intentionally walked |
| 27 | hbp | number of times hit by pitch |
| 28 | sacrhit | number of sacrifice hits |
| 29 | sacrfly | number of sacrifice flies |
| 30 | gidplay | grounded into double play |

**books.csv** Data from homeowner's survey to estimate total number of books, used in Chapter 5.

| Column | Name | Value |
|---|---|---|
| 1 | shelf | shelf number |
| 2 | Mi | number of books on shelf |
| 3 | booknumber | number of the book selected |
| 4 | purchase | purchase cost of book |
| 5 | replace | replacement cost of book |

**census1920.csv** Population sizes for each state, from the 1920 U.S. census. The data set contains only the 48 states and excludes Washington D.C., Puerto Rico, and U.S. territories (these areas were not allowed to have voting representatives in Congress). Source: U.S. Bureau of the Census (1921).

| Column | Name | Value |
|---|---|---|
| 1 | state | state name |
| 2 | population | state population in 1920 census |

**census2010.csv**    Population sizes for each state, from the 2010 U.S. census. Source: U.S. Census Bureau (2019). The data set contains only the 50 states and excludes the areas that, as of 2010, were not allowed to have voting representatives in Congress: Washington D.C., Puerto Rico, and U.S. territories.

| Column | Name | Value |
|---|---|---|
| 1 | state | state name |
| 2 | population | state population in 2010 census |

**cherry.csv**    Data for a sample of 31 cherry trees. Source: Hand et al. (1994).

| Column | Name | Value |
|---|---|---|
| 1 | diameter | Diameter of tree (inches) |
| 2 | height | Height of tree (feet) |
| 3 | volume | Timber volume of tree (cubic feet) |

**classes.csv**    Population sizes for 15 classes, used in Chapter 6 of SDA to illustrate unequal-probability sampling.

| Column | Name | Value |
|---|---|---|
| 1 | class | Class ID number |
| 2 | class_size | Number of students in class |

**classpps.csv**    Two-stage unequal-probability sample without replacement from the population of classes in `classes.csv`.

| Column | Name | Value |
|---|---|---|
| 1 | class | Class ID number |
| 2 | class_size | Number of students in class |
| 3 | finalweight | Sampling weight for student |
| 4 | hours | Number of hours spent studying statistics |

**classppsjp.csv**    Joint inclusion probabilities for unequal-probability sample without replacement from the population of classes in `classes.csv`.

| Column | Name | Value |
|---|---|---|
| 1 | class | Class ID number |
| 2 | class_size | Number of students in class |
| 3 | SelectionProb | Probability of being included in sample, $\pi_i$ |
| 4 | SamplingWeight | Sampling weight $w_i = 1/\pi_i$ |
| 5–9 | JtProb_1–JtProb_5 | Columns of joint inclusion probabilities, $\pi_{ik}$ |

**college.csv**    Selected variables from the U.S. Department of Education College Scorecard Data (version updated on June 1, 2020). Source: U.S. Department of Education (2020), downloaded on August 25, 2020. Some of the variables in `college.csv` have been calculated from other variables in the original source; these have been given new variable names that are not found in the data dictionary at `https://collegescorecard.ed.gov/data/documentation/`.

This data set is made available for pedagogical purposes only. Anyone wishing to draw conclusions from College Scorecard data should obtain the full data set from the Department of Education. The original data set has 1,925 variables and includes institutions (such as those that do not grant undergraduate degrees) that are not in the file `college.csv`.

The data set `college.csv` includes institutions in the original data set that: (1) are located in the 50 states plus the District of Columbia, (2) contain information on average net price (NPT4), (3) are predominantly Bachelor's degree-granting, (4) were currently operating as of June 2020, (5) are not private for-profit institutions or "global" campuses, (6) have Carnegie size classification (variable *ccsizset*) between 6 and 17 and Carnegie basic classification (variable *ccbasic*) between 14 and 22 (these offer Bachelor's degrees), (7) enroll first-time students, and (8) are not U.S. Service Academies.

For all variables, missing data are coded as −99.

| Column | Name | Value |
|---|---|---|
| 1 | unitid | Unit identification number |
| 2 | instnm | Institution name (character, length 81) |
| 3 | city | City (character, length 24) |
| 4 | stabbr | State abbreviation (character, length 2) |
| 5 | highdeg | Highest degree awarded |
|  |  | 3 = Bachelor's degree, 4 = Graduate degree |
| 6 | control | Control (ownership) of institution |
|  |  | 1 = Public, 2 = Private nonprofit |
| 7 | region | Region where institution is located |
|  |  | 1 New England (CT, ME, MA, NH, RI, VT) |
|  |  | 2 Mid East (DE, DC, MD, NJ, NY, PA) |
|  |  | 3 Great Lakes (IL, IN, MI, OH, WI) |
|  |  | 4 Plains (IA, KS, MN, MO, NE, ND, SD) |
|  |  | 5 Southeast (AL, AR, FL, GA, KY, LA, MS, NC, SC, TN, VA, WV) |
|  |  | 6 Southwest (AZ, NM, OK, TX) |
|  |  | 7 Rocky Mountains (CO, ID, MT, UT, WY) |
|  |  | 8 Far West (AK, CA, HI, NV, OR, WA) |
| 8 | locale | Locale of institution |
|  |  | 11 City: Large (population of 250,000 or more) |
|  |  | 12 City: Midsize (population of at least 100,000 but less than 250,000) |
|  |  | 13 City: Small (population less than 100,000) |
|  |  | 21 Suburb: Large (outside principal city, in urbanized area with population of 250,000 or more) |
|  |  | 22 Suburb: Midsize (outside principal city, in urbanized area with population of at least 100,000 but less than 250,000) |

| Column | Name | Value |
|--------|------|-------|
| **college.csv (continued)** | | |
| | | 23 Suburb: Small (outside principal city, in urbanized area with population less than 100,000) |
| | | 31 Town: Fringe (in urban cluster up to 10 miles from an urbanized area) |
| | | 32 Town: Distant (in urban cluster more than 10 miles and up to 35 miles from an urbanized area) |
| | | 33 Town: Remote (in urban cluster more than 35 miles from an urbanized area) |
| | | 41 Rural: Fringe (rural territory up to 5 miles from an urbanized area or up to 2.5 miles from an urban cluster) |
| | | 42 Rural: Distant (rural territory more than 5 miles but up to 25 miles from an urbanized area or more than 2.5 and up to 10 miles from an urban cluster) |
| | | 43 Rural: Remote (rural territory more than 25 miles from an urbanized area and more than 10 miles from an urban cluster) |
| 9 | ccbasic | Carnegie basic classification |
| | | 15 Doctoral Universities: Very High Research Activity |
| | | 16 Doctoral Universities: High Research Activity |
| | | 17 Doctoral/Professional Universities |
| | | 18 Master's Colleges & Universities: Larger Programs |
| | | 19 Master's Colleges & Universities: Medium Programs |
| | | 20 Master's Colleges & Universities: Small Programs |
| | | 21 Baccalaureate Colleges: Arts & Sciences Focus |
| | | 22 Baccalaureate Colleges: Diverse Fields |
| 10 | ccsizset | Carnegie classification, size and setting |
| | | 6 Four-year, very small, primarily nonresidential |
| | | 7 Four-year, very small, primarily residential |
| | | 8 Four-year, very small, highly residential |
| | | 9 Four-year, small, primarily nonresidential |
| | | 10 Four-year, small, primarily residential |
| | | 11 Four-year, small, highly residential |
| | | 12 Four-year, medium, primarily nonresidential |
| | | 13 Four-year, medium, primarily residential |
| | | 14 Four-year, medium, highly residential |
| | | 15 Four-year, large, primarily nonresidential |
| | | 16 Four-year, large, primarily residential |
| | | 17 Four-year, large, highly residential |
| 11 | hbcu | Historically black college or university, $1 =$ yes, $0 =$ no |
| 12 | openadmp | Does the college have an open admissions policy, that is, does it accept any students that apply or have minimal requirements for admission? $1 =$ yes, $0 =$ no |
| 13 | adm_rate | Fall admissions rate, defined as the number of admitted undergraduates divided by the number of undergraduates who applied |
| 14 | sat_avg | Average SAT score (or equivalent) for admitted students |
| 15 | ugds | Number of number of degree-seeking undergraduate students enrolled in the fall term |

| | college.csv (continued) | |
|---|---|---|
| Column | Name | Value |
| 16 | ugds_men | Proportion of *ugds* who are men |
| 17 | ugds_women | Proportion of *ugds* who are women |
| 18 | ugds_white | Proportion of *ugds* who are white (based on self-reports) |
| 19 | ugds_black | Proportion of *ugds* who are black/African American (based on self-reports) |
| 20 | ugds_hisp | Proportion of *ugds* who are Hispanic (based on self-reports) |
| 21 | ugds_asian | Proportion of *ugds* who are Asian (based on self-reports) |
| 22 | ugds_other | Proportion of *ugds* who have other race/ethnicity (created from other categories on original data file; race/ethnicity proportions sum to 1) |
| 23 | npt4 | Average net price of attendance, derived from the full cost of attendance (including tuition and fees, books and supplies, and living expenses) minus federal, state, and institutional grant/scholarship aid, for full-time, first-time undergraduate Title IV-receiving students. NPT4 created from scorecard data variables NPT4_PUB if public institution and NPT4_PRIV if private |
| 24 | tuitionfee_in | In-state tuition and fees |
| 25 | tuitionfee_out | Out-of-state tuition and fees |
| 26 | avgfacsal | Average faculty salary per month |
| 27 | pftfac | Proportion of faculty that is full-time |
| 28 | c150_4 | Proportion of first-year, full-time students who complete their degree within 150% of the expected time to complete; for most institutions, this is the proportion of students who receive a degree within 6 years |
| 29 | grads | Number of graduate students |

**collegerg.csv**   Five replicate SRSs from the set of public colleges and universities (having *control* = 1) in `college.csv`. Columns 1–29 are as in `college.csv`, with additional columns 30–32 listed below. Note that the selection probabilities and sampling weights are for the separate replicate samples, so that the weights for each replicate sample sum to the population size 500.

| Column | Name | Value |
|---|---|---|
| 30 | selectionprob | Selection probability for each replicate sample |
| 31 | samplingweight | Sampling weight for each replicate sample |
| 32 | repgroup | Replicate group number |

**collshr.csv**   Probability-proportional-to-size sample of size 10 from the stratum of small, highly residential colleges (having *ccsizeset* = 11) in `college.csv`. Columns 1–29 are as in `college.csv`, with additional columns 30–34 listed below.

| Column | Name | Value |
|--------|------|-------|
| 30 | mathfac | Number of mathematics faculty |
| 31 | psychfac | Number of psychology faculty |
| 32 | biolfac | Number of biology faculty |
| 33 | psii | Selection probability, $= ugds/$(sum of $ugds$ for stratum) |
| 34 | wt | Sampling weight $= 1/(10\psi_i)$ |

**coots.csv**  Selected information on egg size, from a larger study by Arnold (1991). Data provided courtesy of Todd Arnold. Not all observations are used for this data set, so results may not agree with those in Arnold (1991).

| Column | Name | Value |
|--------|------|-------|
| 1 | clutch | clutch number from which eggs were subsampled. |
| 2 | csize | number of eggs in clutch ($M_i$) |
| 3 | length | length of egg (mm) |
| 4 | breadth | maximum breadth of egg (mm) |
| 5 | volume | calculated as 0.000507*length * breadth$^2$ (mm$^3$) |
| 6 | tmt | $= 1$ if received supplemental feeding, 0 otherwise |

**counties.csv**  Data (from 1990) from an SRS of 100 of the 3141 counties in the United States. Missing values are coded as $-99$. Source: U.S. Census Bureau (1994).

| Column | Name | Value |
|--------|------|-------|
| 1 | RN | random number used to select the county |
| 2 | state | state abbreviation |
| 3 | county | county name |
| 4 | landarea | land area, 1990 (square miles) |
| 5 | totpop | total number of persons, 1992 |
| 6 | physician | active non-Federal physicians on Jan. 1, 1990 |
| 7 | enroll | school enrollment in elementary or high school, 1990 |
| 8 | percpub | percent of school enrollment in public schools |
| 9 | civlabor | civilian labor force, 1991 |
| 10 | unemp | number unemployed, 1991 |
| 11 | farmpop | farm population, 1990 |
| 12 | numfarm | number of farms, 1987 |
| 13 | farmacre | acreage in farms, 1987 |
| 14 | fedgrant | total expenditures in federal funds and grants, 1992 (millions of dollars) |
| 15 | fedciv | civilians employed by federal government, 1990 |
| 16 | milit | military personnel, 1990 |
| 17 | veterans | number of veterans, 1990 |
| 18 | percviet | percent of veterans from Vietnam era, 1990 |

**crimes.csv**  Data from selected variables in a simple random sample of 5,000 records from the 7,048,107 records with dates between 2001 and 2019 in the City of Chicago database "Crimes—2001 to Present." This file was downloaded on August 11, 2020 from https://data.cityofchicago.org/. These data are provided for pedagogical purposes only. Anyone wishing to publish analyses of Chicago crime data should obtain the most recent data from `https://data.cityofchicago.org/`. For a list and map of Community Areas, see `https://www.chicago.gov/city/en/depts/dgs/supp_info/citywide_maps.html`.

| Column | Name | Value |
|---|---|---|
| 1 | year | Year in which crime occurred (between 2001 and 2019) |
| 2 | crimetype | Type of crime, determined from detailed crime description in database |
| | | homicide = homicide, sexualasslt = sexual assault, robbery = robbery, aggasslt = aggravated assault, burglary = burglary, mvtheft = motor vehicle theft, idtheft = identity theft, theft = other type of theft, arson = arson, simpleasslt = simple assault (assaults that are not aggravated), threat = threat or harassment, fraud = fraud, weapon = weapons violation, trespass = trespassing, narcotics = narcotics or liquor law violation, vandalism = vandalism, other = other |
| 3 | violent | = 1 if violent crime, 0 otherwise |
| 4 | arrest | = 1 if an arrest was made, 0 otherwise |
| 5 | domestic | = 1 if crime was domestic-related as defined by the Illinois Domestic Violence Act, 0 otherwise |
| 6 | commarea | Number of the Community Area in Chicago where the crime occurred |
| 7 | location | Type of location where crime occurred (e.g., street, apartment) |

**deadtrees.csv** Number of dead trees recorded by photograph and field count for a (fictional) SRS of 25 plots taken from a population of 100 plots.

| Column | Name | Value |
|---|---|---|
| 1 | photo | Number of dead trees in plot from photograph |
| 2 | field | Number of dead trees in plot from field observation |

**divorce.csv** Data from a sample of divorce records for states in the Divorce Registration Area. Source: National Center for Health Statistics (1987).

| Column | Name | Value |
|---|---|---|
| 1 | state | state name (character variable) |
| 2 | abbrev | state abbreviation (character variable) |
| 3 | samprate | sampling rate for state |
| 4 | numrecs | number of records sampled in state |
| 5 | hsblt20 | number of records in sample with husband's age < 20 |
| 6 | hsb20to24 | number of records with $20 \leq$ husband's age $\leq 24$ |
| 7 | hsb25to29 | number of records with $25 \leq$ husband's age $\leq 29$ |
| 8 | hsb30to34 | number of records with $30 \leq$ husband's age $\leq 34$ |
| 9 | hsb35to39 | number of records with $35 \leq$ husband's age $\leq 39$ |
| 10 | hsb40to44 | number of records with $40 \leq$ husband's age $\leq 44$ |

**divorce.csv (continued)**

| Column | Name | Value |
|--------|------|-------|
| 11 | hsb45to49 | number of records with $45 \leq$ husband's age $\leq 49$ |
| 12 | hsbge50 | number of records with husband's age $\geq 50$ |
| 13 | wflt20 | number of records with wife's age $< 20$ |
| 14 | wf20to24 | number of records with $20 \leq$ wife's age $\leq 24$ |
| 15 | wf25to29 | number of records with $25 \leq$ wife's age $\leq 29$ |
| 16 | wf30to34 | number of records with $30 \leq$ wife's age $\leq 34$ |
| 17 | wf35to39 | number of records with $35 \leq$ wife's age $\leq 39$ |
| 18 | wf40to44 | number of records with $40 \leq$ wife's age $\leq 44$ |
| 19 | wf45to49 | number of records with $45 \leq$ wife's age $\leq 49$ |
| 20 | wfge50 | number of records with wife's age $\geq 50$ |

**gini.csv**   Data from the population of districts for the 1921 Italian general census. Source: Gini and Galvani (1929, pp. 73–78).

| Column | Name | Value |
|--------|------|-------|
| 1 | id | ID number |
| 2 | district | District name |
| 3 | birth_rate | Births per 1,000 population |
| 4 | death_rate | Deaths per 1,000 population |
| 5 | marriage_rate | Marriages per 1,000 population |
| 6 | agricultural_pop | Percentage of males over 10 years old who work in agriculture |
| 7 | urban_population | Percentage of population in urban areas |
| 8 | income | Average income |
| 9 | altitude | Average altitude above sea level (meters) |
| 10 | pop_density | Number of inhabitants per square kilometer |
| 11 | natural_growth | Rate of average increase of the population |
| 12 | population | Population of area |
| 13 | area | Land area (square kilometers) |
| 14 | in_GG_sample | $= 1$ if in the purposive sample selected by Gini and Galvani; 0 otherwise |

**golfsrs.csv**   A simple random sample of 120 golf courses, taken from the population on the website ww2.golfcourse.com on August 5, 1998. Missing data in the .csv file are denoted by blanks.

| Column | Name | Value |
|--------|------|-------|
| 1 | RN | random number used to select golf course for sample |
| 2 | state | state name |
| 3 | holes | number of holes |
| 4 | type | type of course: priv = private, semi = semi-private, pub = public, mili = military, resort |
| 5 | yearblt | year course was built |
| 6 | wkday18 | greens fee for 18 holes during week |
| 7 | wkday9 | greens fee for 9 holes during week |
| 8 | wkend18 | greens fee for 18 holes on weekend |

| golfsrs.csv (continued) | | |
|---|---|---|
| Column | Name | Value |
| 9 | wkend9 | greens fee for 9 holes on weekend |
| 10 | backtee | back tee yardage |
| 11 | rating | course rating |
| 12 | par | par for course |
| 13 | cart18 | golf cart rental fee for 18 holes |
| 14 | cart9 | golf cart rental fee for 9 holes |
| 15 | caddy | Are caddies available? (y or n) |
| 16 | pro | Is a golf pro available? (y or n) |

**gpa.csv**   GPA data from Chapter 5 of SDA.

| Column | Name | Value |
|---|---|---|
| 1 | suite | Suite (psu) identifier |
| 2 | gpa | Grade point average of person in suite |
| 3 | wt | Sampling weight, = 20 for every observation |

**healthjournals.csv**   Randomization and statistical inference practices in a stratified random sample of 196 public health articles. The data, provided courtesy of Dr. Matt Hayat, are discussed in Hayat and Knapp (2017). The variables provided in `healthjournals.csv` are a subset of the variables collected by the authors.

| Column | Name | Value |
|---|---|---|
| 1 | journal | Journal that published the article<br>AJPH = *American Journal of Public Health*<br>AJPM = *American Journal of Preventive Medicine*<br>PM = *Preventive Medicine* |
| 2 | NumAuthors | Number of authors |
| 3 | RandomSel | = "Yes" if data in the article were from a randomly selected (probability) sample; "No" otherwise |
| 4 | RandomAssn | = "Yes" if study subjects for the article were randomly assigned to treatment groups; "No" otherwise |
| 5 | ConfInt | = "Yes" if a confidence interval appeared in the article's main text, tables, or figures; "No" otherwise |
| 6 | HypTest | = "Yes" if a p-value or significance test appeared in the article's main text, tables, or figures; "No" otherwise |
| 7 | Asterisks | = "Yes" if asterisks were used to represent p-value ranges; "No" otherwise |

**htcdf.csv**   Empirical distribution function and empirical probability mass function of data in `htpop.csv`.

| Column | Name | Value |
|---|---|---|
| 1 | height | height value, cm |
| 2 | frequency | number of times height value in column 1 occurs in population |
| 3 | epmf | empirical probability mass function |
| 4 | ecdf | empirical distribution function |

**htpop.csv**   Height and gender of 2,000 persons in an artificial population.

| Column | Name | Value |
|---|---|---|
| 1 | height | height of person, cm |
| 2 | gender | M=male, F=female |

**htsrs.csv**   Height and gender for a SRS of 200 persons, taken from `htpop.csv`.

| Column | Name | Value |
|---|---|---|
| 1 | rn | random number used to select unit |
| 2 | height | height of person, cm |
| 3 | gender | M=male, F=female |

**htstrat.csv**   Height and gender for a stratified random sample of 160 women and 40 men, taken from `htpop.csv`. The columns and names are as in `htsrs.csv`.

**hunting.csv**   Population and sample sizes for the poststrata used for the Sunday hunting survey. Source: Virginia Polytechnic and State University/Responsive Management (2006).

| Column | Name | Value |
|---|---|---|
| 1 | region | Region of state (East, Central, West) |
| 2 | gender | Gender (female, male) |
| 3 | age | Age group (16-24, 25-34, 35-44, 45-54, 55-64, 65+) |
| 4 | popsize | Population size in poststratum from the 2000 U.S. census |
| 5 | sampsize | Sample size in poststratum |

**impute.csv**   Small artificial data set used to illustrate imputation methods. Missing values are denoted by $-99$.

| Column | Name | Value |
|---|---|---|
| 1 | person | identification number for person |
| 2 | age | age in years |
| 3 | gender | M=male, F=female |
| 4 | education | number of years of education |
| 5 | crime | = 1 if victim of any crime, 0 otherwise |
| 6 | violcrime | = 1 if victim of violent crime, 0 otherwise |

**integerwt.csv**   Artificial population of 2,000 observations.

| Column | Name | Value |
|---|---|---|
| 1 | stratum | Stratum number |
| 2 | y | $y$ value of observation |

**intellonline.csv** Data from the online (Mechanical Turk) survey. Source: Heck et al. (2018). The data were downloaded from `https://journals.plos.org/plosone/article?id=10.1371/journal.pone.0200103` on February 8, 2020; the variables extracted from the full data set are provided here for educational purposes only.

| Column | Name | Value |
|---|---|---|
| 1 | int | Response to question about agreement with the statement "I am more intelligent than the average person." 1 = Strongly Agree; 2 = Mostly Agree; 3 = Mostly Disagree; 4 = Strongly Disagree; 5 = Don't Know or Not Sure |
| 2 | region | Census region of respondent (character variable, length 10): Northeast, South, Midwest, West |
| 3 | sex | Sex (character variable, length 8): Male, Female |
| 4 | race | Race (character variable, length 18): White, African American, Asian American, Hispanic American, Another origin |
| 5 | age | Age, years |
| 6 | income | Household income level (character variable, length 8): <$40k, $40–80k, or >$80k |
| 7 | education | Highest education level attained (character variable, length 12): No College, Some College, College Grad, Grad School |
| 8 | postwt | Relative weight, obtained by poststratifying to demographic proportions in the 2010 U.S. Census. The weights are normed so that they sum to 750. |

**intelltel.csv** Data from the telephone survey studied by Heck et al. (2018). The data were downloaded from `https://journals.plos.org/plosone/article?id=10.1371/journal.pone.0200103` and are provided here for educational purposes only. The variables are the same as in `intellonline.csv`.

**intellwts.csv** Relative weights for demographic groups in `intellonline.csv` and `intelltel.csv` (Heck et al., 2018). Each sample was weighted using the 2010 U.S. Census demographics for sex (male, female), age ($< 44$, $\geq 44$), and race/ethnicity (white, nonwhite). The table entries give the weights for each of these eight demographic groups.

| Column | Name | Value |
|---|---|---|
| 1 | sex | Sex |
| 2 | agegroup | Age group: Young = (age less than 44), Old = (age greater than or equal to 44) |
| 3 | race | Race: White or Nonwhite |
| 4 | tel_n | Number of telephone survey respondents in the sex/agegroup/race class |

| Column | Name | Value |
|---|---|---|
| | **intellwts.csv (continued)** | |
| Column | Name | Value |
| 5 | online_n | Number of online survey respondents in the sex/agegroup/race class |
| 6 | tel_wgt | Relative weight for each respondent to the telephone survey in this sex/agegroup/race class |
| 7 | online_wgt | Relative weight for each respondent to the telephone survey in this sex/agegroup/race class |

**ipums.csv** Data extracted from the 1980 Census Integrated Public Use Microdata Series, using the "Small Sample Density" option in the data extract tool, on September 17, 2008. The stratum and psu variables were constructed for use in the book exercises. Data analyses on this file do NOT give valid results for inference to the 1980 U.S. population. Source: Ruggles et al. (2004).

| Column | Name | Value |
|---|---|---|
| 1 | stratum | stratum number (1–9) |
| 2 | psu | psu number (1–90) |
| 3 | inctot | total personal income (dollars), topcoded at $75,000 |
| 4 | age | age, with range 15–90 |
| 5 | sex | 1 = Male, 2 = Female |
| 6 | race | 1 = White, 2 = Black, 3 = American Indian or Alaska Native, 4 = Asian or Pacific Islander, 5 = Other Race |
| 7 | hispanic | 0 = Not Hispanic, 1 = Hispanic |
| 8 | marstat | Marital Status: 1 = Married, 2 = Separated, 3 = Divorced, 4 = Widowed, 5 = Never married/single |
| 9 | ownershg | Ownership of housing unit: 0 = Not Applicable (N/A), 1 = Owned or being bought, 2 = Rents |
| 10 | yrsusa | Number of years a foreign-born person has lived in the U.S.: 0= N/A, 1= 0–5 years, 2= 6–10 years, 3= 11–15 years, 4= 16–20 years, 5= 21+ years |
| 11 | school | Is person in school? 0 = N/A, 1 = No, not in school, 2 = Yes, in school |
| 12 | educrec | Educational Attainment: 1= None or preschool, 2= Grade 1, 2, 3, or 4, 3= Grade 5, 6, 7, or 8, 4= Grade 9, 5= Grade 10, 6= Grade 11, 7= Grade 12, 8= 1 to 3 years of college, 9= 4+ years of college |
| 13 | labforce | In labor force? 0 = Not Applicable, 1 = No, 2 = Yes |
| 14 | classwk | class of worker: 0=Not applicable, 13= Self-employed, not incorporated, 14= Self-employed, incorporated, 22= Wage/salary, private, 25= Federal government employee, 27= State government employee, 28= Local government employee, 29= Unpaid family worker |
| 15 | vetstat | Veteran Status 0 = Not Applicable, 1 = No Service, 2 = Yes |

**journal.csv**  Types of sampling used for articles in a sample of journals. Source: Jacoby and Handlin (1991).

Note that columns 2 and 3 do not always sum to column 1; for some articles, the investigators could not determine which type of sampling was used. When working with these data, you may wish to create a fourth column, "indeterminate," which equals column1 − (column2 + column3).

| Column | Name | Value |
|---|---|---|
| 1 | numemp | number of articles in 1988 that used sampling |
| 2 | prob | number of articles that used probability sampling |
| 3 | nonprob | number of articles that used non-probability sampling |

**measles.csv**  Roberts et al. (1995) reported on the results of a survey of parents whose children had not been immunized against measles during a recent campaign to immunize all children in the first five years of secondary school. The original data were unavailable; univariate and multivariate summary statistics from these artificial data, however, are consistent with those in the paper. All variables are coded as 1 for yes, 0 for no, and 9 for no answer. A parent who refused consent (variable 4) was asked why, with· responses in variables 5 through 10. If a response in variables 5 through 10 was checked, it was assigned value 1; otherwise, it was assigned value 0. A parent could give more than one reason for not having the child immunized.

| Column | Name | Value |
|---|---|---|
| 1 | school | school attended by child |
| 2 | form | Parent received consent form |
| 3 | returnf | Parent returned consent form |
| 4 | consent | Parent gave consent for measles immunization |
| 5 | hadmeas | Child had already had measles |
| 6 | previmm | Child had been immunized against measles |
| 7 | sideeff | Parent concerned about side effects |
| 8 | gp | Parent wanted general practitioner (GP) to give vaccine |
| 9 | noshot | Child did not want injection |
| 10 | notser | Parent thought measles was not a serious illness |
| 11 | gpadv | GP advised that vaccine was not needed |
| 12 | Mitotal | Population size in school |
| 13 | mi | Sample size in school |

**mysteries.csv**  Data from a stratified random sample of books nominated for the Edgar® awards for Best Novel and Best First Novel. The sample was drawn from the population listing of 655 books at http://theedgars.com/awards/ on August 14, 2020.

| Column | Name | Value |
|---|---|---|
| 1 | stratum | Stratum number, from 1 to 12, computed from the stratification variables in columns 2–4 |
| 2 | time | Time period in which award was given: 1 = 1946–1980, 2 = 1981–2000, 3 = 2001–2020 |
| 3 | category | Award category (character variable, length 16): Best Novel, or Best First Novel |

| mysteries.csv (continued) | | |
| --- | --- | --- |
| **Column** | **Name** | **Value** |
| 4 | winner | = 1 if book won the award that year, = 0 if book was nominated but did not win award |
| 5 | popsize | Number of population books in stratum ($= N_h$) |
| 6 | sampsize | Number of sampled books in stratum ($= n_h$) |
| 7 | obtained | = 1 if book was obtained (responded) in original sample, = 2 if book was obtained in phase II subsample of nonrespondents, = 0 if not obtained |
| 8 | p1weight | Weight for phase I sample, calculated as $N_h/n_h$; use for exercises in Chapters 1–11 of SDA |
| 9 | p2weight | Final weight for phase II sample; use for exercises in Chapter 12 of SDA and analyses involving variables *victims* and *firearm* |
| 10 | genre | Genre of book (character variable, length 11). Values "private eye" (protagonist is a private detective), "procedural" (a detailed, step-by-step analysis of how the crime is solved, using the skills of the detective), or "suspense" (the protagonist is at the center of action or is involved in espionage, but is not a professional detective) |
| 11 | historical | = 1 if the main action in the book takes place at least 20 years before the book's publication date, = 0 if book action is within 20 years of the publication date |
| 12 | urban | = 1 if the main action in the book takes place primarily in urban areas, = 0 otherwise |
| 13 | authorgender | Gender of author (character variable, length 1) = "F" if author is female, "M" if author is male |
| 14 | fdetect | Number of female detectives (or protagonists, if book has no detective) in book |
| 15 | mdetect | Number of male detectives (or protagonists, if book has no detective) in book |
| 16 | victims | Number of murder victims in book (missing value set to −9 if *obtained* = 0) |
| 17 | firearm | Number of murders committed with firearms in book (missing value set to −9 if *obtained* = 0) |

**nhanes.csv**    Selected variables from the 2015–2016 National Health and Nutrition Examination Survey (NHANES). Source: Centers for Disease Control and Prevention (2017). This data set is provided for educational purposes only. Anyone wishing to publish or use results from analyses of NHANES data should obtain the data files directly from the source.

The data files merged to create nhanes.csv can be read directly from the SAS transport files DEMO_I.XPT, BMX_I.XPT, TCHOL_I.XPT, and BPX_I.XPT from the NHANES website. Variables 1–23 have the same names as in the SAS transport files.

The blood pressure variables *sbp* and *dbp* were created as follows. In the medical examination, three consecutive blood pressure readings were obtained after participants sat quietly for 5 minutes, and the maximum inflation level was determined. A fourth measurement was conducted for some persons who had an incomplete or interrupted blood pressure reading.

The variables *sbp* and *dbp* were calculated by discarding the first blood pressure reading and calculating the average of the remaining valid readings. Note that some of the diastolic blood pressure readings are 0.

In the comma-delimited file `nhanes.csv`, missing values are denoted by −9. In the SAS data file, missing values are denoted by a period. In the R data file, missing values are denoted by NA. Note that some of the codes for variables in the table below also denote missing values; for example, the value 7 for *dmdeduc2* indicates "Refused," and these codes for special types of missing values remain in the SAS and R data files.

| Column | Name | Value |
|---|---|---|
| 1 | sdmvstra | Pseudo-stratum. These are groups of secondary sampling units used for variance estimation on the publicly available data. Pseudo-strata and pseudo-psus are released instead of the actual strata and psus to protect the confidentiality of respondents' information. Use *sdmvstra* as the variable defining the strata. |
| 2 | sdmvpsu | Pseudo-psu. Use *sdmvpsu* as the primary sampling unit (psu). There are two pseudo-psus per pseudo-stratum, numbered 1 and 2. |
| 3 | wtint2yr | Interview weight (use as weight for variables 5–12) |
| 4 | wtmec2yr | Mobile Examination Center weight (use as weight for any analysis involving variables 13–25) |
| 5 | ridstatr | Interview/examination status, = 1 if interviewed only, = 2 if interviewed and had medical examination |
| 6 | ridageyr | Age in years at screening, from 0 to 80. Anyone with age > 80 years is recorded (topcoded) as 80. No values are missing for this variable. |
| 7 | ridagemn | Age in months at screening (reported only for persons with age 24 months or younger at the time of exam, otherwise missing) |
| 8 | riagendr | = 1 if male, 2 if female (no missing values) |
| 9 | ridreth3 | Race/ethnicity code (no missing values)<br>1 = Mexican American<br>2 = Other Hispanic<br>3 = Non-Hispanic White<br>4 = Non-Hispanic Black<br>6 = Non-Hispanic Asian<br>7 = Other Race, Including Multi-Racial |
| 10 | dmdeduc2 | Education level of person interviewed (given for adults age 20+ only)<br>1 = Less than 9th grade<br>2 = 9th to 11th grade (including 12th grade with no diploma)<br>3 = High school graduate (including GED)<br>4 = Some college or associate's degree<br>5 = College graduate or above<br>7 = Refused<br>9 = Don't know |
| 11 | dmdfmsiz | Total number of people in the family. Values 1–6 indicate the number of people is that number; value 7 indicates 7 or more people in family. No missing values. |

### nhanes.csv (continued)

| Column | Name | Value |
|--------|------|-------|
| 12 | indfmpir | Ratio of family income to poverty guideline. A value less than 1 indicates the family is below the poverty threshold. Variable *indfmpir* is a continuous variable where values between 0 and 4.99 indicate the actual poverty ratio. A value of 5 indicates that the ratio of family income to the poverty guideline for that family is 5 or more. |
| 13 | bmxwt | Weight (kg) |
| 14 | bmxht | Standing height (cm) |
| 15 | bmxbmi | Body mass index $(kg/m^2)$, calculated as $bmxwt/(bmxht/100)^2$ |
| 16 | bmxwaist | Waist circumference (cm) |
| 17 | bmxleg | Upper leg length (cm) |
| 18 | bmxarml | Upper arm length (cm) |
| 19 | bmxarmc | Upper arm circumference (cm) |
| 20 | bmdavsad | Average sagittal abdominal diameter (SAD, the distance from the small of the back to the upper abdomen), in cm. Calculated by averaging the SAD readings on the person (up to four). |
| 21 | lbxtc | Serum total cholesterol (mg/dL) |
| 22 | bpxpls | 60-second pulse |
| 23 | sbp | Average systolic blood pressure (mm Hg) |
| 24 | dbp | Average diastolic blood pressure (mm Hg) |
| 25 | bpread | Number of blood pressure readings |

**nybight.csv**    Data collected in the New York Bight for June 1974 and June 1975. Two of the original strata were combined because of insufficient sample sizes. For variable *catchwt*, weights less than 0.5 were recorded as 0.5 kg. Source: Wilk et al. (1977).

| Column | Name | Value |
|--------|------|-------|
| 1 | year | year of data collection, 1974 or 1975 |
| 2 | stratum | stratum membership, based on depth |
| 3 | catchnum | number of fish caught during trawl |
| 4 | catchwt | total weight (kg) of fish caught during trawl |
| 5 | numspp | number of species of fish caught during trawl |
| 6 | depth | depth of station (m) |
| 7 | temp | surface temperature (degrees C), missing $= -99$ |

**otters.csv**    Data on number of holts (dens) in Shetland, U.K., used in Kruuk et al. (1989). Data courtesy of Hans Kruuk.

| Column | Name | Value |
|--------|------|-------|
| 1 | section | section of coastline |
| 2 | habitat | type of habitat (stratum) |
| 3 | holts | number of holts (dens) |

**ozone.csv**    Hourly ozone readings (parts per billion, ppb) from a site in Monterey County, California, for 2018 and 2019. Source: `https://aqs.epa.gov/aqsweb/airdata/download_files.html#Raw`, accessed November 19, 2020. Missing values are denoted by $-9$.

| Column | Name | Value |
|---|---|---|
| 1 | year | year of reading (2018 or 2019) |
| 2 | month | month of reading (1–12) |
| 3 | day | day of reading (1–31) |
| 4 | hr0 | ozone reading (ppb) at 0:00 local time |
| 5 | hr1 | ozone reading (ppb) at 1:00 local time |
| ⋮ | ⋮ | ⋮ |
| 27 | hr23 | ozone reading (ppb) at 23:00 local time |

**pitcount.csv**  Fictional data from a fictional point-in-time (PIT) survey taken to estimate the number of persons experiencing homelessness.

| Column | Name | Value |
|---|---|---|
| 1 | strat | Stratum number (from 1 to 8) |
| 2 | division | Geographic division, used to form strata |
| 3 | density | Expected density of persons experiencing homelessness (character variable, with values High or Low) |
| 4 | popsize | $= N_h$, the number of areas in the population for stratum $h$ |
| 5 | sampsize | $= n_h$, the number of areas in the sample for stratum $h$ |
| 6 | areawt | $= N_h/n_h$, the sampling weight for the area |
| 7 | y | Number of persons experiencing unsheltered homelessness found in the area during the PIT count |

**profresp.csv**  The data described in Zhang et al. (2020) were downloaded from http://doi.org/10.3886/E109021V1 on January 22, 2020, from file survey4.rds. The data set profresp.csv contains selected variables from the set of 2,407 respondents who completed the survey and provided information on the demographic variables and the information needed to calculate "professional respondent" status. The full data set survey4.rds contains numerous additional questions about behavior that are not included here, as well as the data from the partially completed surveys. The website also contains data for three other online panel surveys. Because profresp.csv is a subset of the full data, statistics calculated from it may differ from those in Zhang et al. (2020).

Missing values are denoted by −9.

| Column | Name | Value |
|---|---|---|
| 1 | prof_cat | Level of professionalism 1 = novice, 2 = average, 3 = professional |
| 2 | panelnum | Number of panels respondent has belonged to. A response between 1 and 6 means that the person has belonged to that number of panels; 7 means 7 or more. |
| 3 | survnum_cat | How many Internet surveys have you completed before this one? 1 = This is my first one, 2 = 1–5, 3 = 6–10, 4 = 11–15, 5 = 16–20, 6 = 21–30, 7 = More than 30 |
| 4 | panelq1 | Are you a member of any online survey panels besides this one? 1 = yes, 2 = no |
| 5 | panelq2 | To how many other online panels do you belong? |

| Column | Name | Value |
|--------|------|-------|
| | | **profresp.csv (continued)** |

| Column | Name | Value |
|--------|------|-------|
| | | 1 = None, 2 = 1 other panel, 3 = 2 others, 4 = 3 others, 5 = 4 others, 6 = 5 others, 7 = 6 others or more. This question has a missing value if *panelq1* = 2. If you want to estimate how many panels a respondent belongs to, create a new variable *numpanel* that equals *panelq2* if *panelq2* is not missing and equals 1 if *panelq1* = 2. |
| 6 | age4cat | Age category. 1 = 18 to 34, 2 = 35 to 49, 3 = 50 to 64, 4 = 65 and over |
| 7 | edu3cat | Education category. 1 = high school or less, 2 = some college or associates' degree, 3 = college graduate or higher |
| 8 | gender | Gender: 1 = male, 2 = female |
| 9 | non_white | 1 = race is non-white, 0 = race is white |
| 10 | motive | Which best describes your main reason for joining on-line survey panels? 1 = I want my voice to be heard, 2 = Completing surveys is fun, 3 = To earn money, 4 = Other (Please specify) |
| 11 | freq_q1 | During the PAST 12 MONTHS, how many times have you seen a doctor or other health care professional about your own health? Response is number between 0 and 999. |
| 12 | freq_q2 | During the PAST MONTH, how many days have you felt you did not get enough rest or sleep? |
| 13 | freq_q3 | During the PAST MONTH, how many times have you eaten in restaurants? Please include both full-service and fast food restaurants. |
| 14 | freq_q4 | During the PAST MONTH, how many times have you shopped in a grocery store? If you shopped at more than one grocery store on a single trip, please count them separately. |
| 15 | freq_q5 | During the PAST 2 YEARS, how many overnight trips have you taken? |

**profrespacs.csv**   Population estimates from the 2011 American Community Survey (ACS) for age/gender/education categories measured in `profresp.csv` (Zhang et al., 2020). Note that *age3cat* has 3 categories, while the age variable in `profresp.csv` has 4 categories.

| Column | Name | Value |
|--------|------|-------|
| 1 | gender | Gender: 1 = male, 2 = female |
| 2 | age3cat | Age category. 1 = 18 to 34, 2 = 35 to 64, 3 = 65 and over |
| 3 | edu3cat | Education category. 1 = high school or less, 2 = some college or associates' degree, 3 = college graduate or higher |
| 4 | count | Population size from ACS for the gender/age/education level combination |

**radon.csv**   Radon readings for a stratified sample of 1003 homes in Minnesota. Source: Nolan and Speed (2000). The data were downloaded in April 2008 from an earlier version of the website now located at `www.stat.berkeley.edu/users/statlabs/labs.html`.

| Column | Name | Value |
|---|---|---|
| 1 | countyname | County Name |
| 2 | countynum | County Number |
| 3 | sampsize | Sample size in county |
| 4 | popsize | Population size in county |
| 5 | radon | Radon concentration (picocuries per liter) |

**rectlength.csv**   Lengths of rectangles.

| Column | Name | Value |
|---|---|---|
| 1 | rectangle | Rectangle number |
| 2 | length | Rectangle length |

**rnt.csv**   Page from a random number table. Open the `.csv` file in a text editor instead of a spreadsheet, because a spreadsheet strips off the leading zeroes. The columns have format z5.0 in the SAS file, and are character variables in the R file, so that leading zeroes are displayed in those formats.

| Column | Name | Value |
|---|---|---|
| 1 | col1 | Column of 5-digit random numbers |
| 2 | col2 | Column of 5-digit random numbers |
| 3 | col3 | Column of 5-digit random numbers |
| 4 | col4 | Column of 5-digit random numbers |
| 5 | col5 | Column of 5-digit random numbers |
| 6 | col6 | Column of 5-digit random numbers |

**sample70.csv**   All possible simple random samples that can be generated from the population in Example 2.2 of SDA.

| Column | Name | Value |
|---|---|---|
| 1 | sampnum | Sample number |
| 2–5 | u1–u4 | Sampled units in $\mathcal{S}$ |
| 6–9 | y1–y4 | Values of $y_i$ in sample $\mathcal{S}$ |
| 10 | total | Estimated population total |

**santacruz.csv**   The number of seedlings in the sampled psus on Santa Cruz Island, California, in 1992 and 1994. Source: Peart (1994).

| Column | Name | Value |
|---|---|---|
| 1 | tree | Tree number |
| 2 | seed92 | Number of seedlings in 1992 |
| 3 | seed94 | Number of seedlings in 1994 |

**schools.csv**   Math and reading test results from a two-stage cluster sample of tenth-grade students. An SRS of 10 schools was selected from the 75 schools in the population, and then 20 students were sampled from each school. These data are fictional but the summary statistics are consistent with those seen in educational studies.

| Column | Name | Value |
|---|---|---|
| 1 | schoolid | School number (use as cluster variable) |
| 2 | gender | Gender of student (character variable, F = female, M = male) |
| 3 | math | Score on math test |
| 4 | reading | Score on reading test |
| 5 | mathlevel | Category level for math test score:<br>1 if $1 \leq$ math $<= 40$<br>2 if $41 \leq$ math |
| 6 | readlevel | Category level for reading test score:<br>1 if $1 \leq$ read $<= 32$<br>2 if $33 \leq$ read $<= 50$ |
| 7 | Mi | Number of students in school, $M_i$ |
| 8 | finalwt | Weight for student in sample |

**seals.csv**   Data on number of breathing holes found in sampled areas of Svalbard fjords, reconstructed from summary statistics given in Lydersen and Ryg (1991).

| Column | Name | Value |
|---|---|---|
| 1 | zone | zone number for sampled area |
| 2 | holes | number of breathing holes Imjak found in area |

**shapespop.csv**   Population of black and gray squares and circles.

| Column | Name | Value |
|---|---|---|
| 1 | ID | identification number for object |
| 2 | shape | shape of object (square or circle) |
| 3 | color | color of object (gray or black) |
| 4 | area | area of object ($cm^2$) |
| 5 | conv | = 1 if object can be reached through convenience sample, 0 otherwise |

**shorebirds.csv**   Two-phase sample of shorebird nests. These are artificial data constructed from summary statistics given in Bart and Earnst (2002).

| Column | Name | Value |
|---|---|---|
| 1 | plot | Plot number |
| 2 | rapid | Rapid-method count of number of birds in plot |
| 3 | intense | Intensive-method count of number of nests in plot<br>= −9 if the plot is not in the phase II sample |

**sp500.csv**   Companies in the S&P 500® Stock Market Index as of September 15, 2020. Source: Downloaded from `https://fknol.com/list/eps-sp-500-index-companies.php` on September 19, 2020.

| Column | Name | Value |
| --- | --- | --- |
| 1 | Company | Company name (character variable, length 37) |
| 2 | Symbol | Stock symbol (character variable, length 5) |
| 3 | MarketCap | Market capitalization, in billions of U.S. dollars |
| 4 | StockPrice | Price per share of stock |
| 5 | PE_Ratio | Price-to-earnings ratio |
| 6 | EPS | Earnings per share |

**spanish.csv**   Fictional cluster sample of introductory Spanish students.

| Column | Name | Value |
| --- | --- | --- |
| 1 | class | Class number |
| 2 | score | Score on vocabulary test (out of 100) |
| 3 | trip | = 1 if plan a trip to a Spanish-speaking country, 0 otherwise |

**srs30.csv**   An SRS of size 30 taken from an artificial population of size 100.

| Column | Name | Value |
| --- | --- | --- |
| 1 | y | Value of observation |

**ssc.csv**   SRS of 150 members of the Statistical Society of Canada, downloaded from `ssc.ca` in August, 2006.

| Column | Name | Value |
| --- | --- | --- |
| 1 | gender | m = male, f=female |
| 2 | occupation | a = academic, g = government, i = industry, n = not determined |
| 3 | ASA | = 1 if person is member of American Statistical Association, 0 otherwise |

**statepop.csv**   Data from an unequal-probability sample of 100 counties from the 1994 *County and City Data Book* (U.S. Census Bureau, 1994). The sample was selected with probability proportional to population.

| Column | Name | Value |
| --- | --- | --- |
| 1 | county | county name (character variable, length 14) |
| 2 | state | state name (character variable) |
| 3 | landarea | land area of county, 1990 (square miles) |
| 4 | popn | population of county, 1992 |
| 5 | phys | number of physicians, 1990 |
| 6 | farmpop | farm population, 1990 |
| 7 | numfarm | number of farms, 1987 |
| 8 | farmacre | number of acres devoted to farming, 1987 |
| 9 | veterans | number of veterans, 1990 |
| 10 | percviet | percent of veterans from Vietnam era, 1990 |
| 11 | psii | $\psi_i$, probability of selection |
| 12 | wt | sampling weight, $= 1/(100\psi_i)$ |

**statepps.csv**   Number of counties (or county equivalents; Alaska has boroughs, Louisiana has parishes, and some states have independent cities), population estimates for 2019, land area, and water area for the 50 states plus the District of Columbia. Total area for a state can be calculated by summing land area and water area.

Source: Population estimates are from U.S. Census Bureau (2019). Land and water areas are from U.S. Census Bureau (2012).

| Column | Name | Value |
|---|---|---|
| 1 | state | state name (character variable, length 20) |
| 2 | counties | number of counties or county equivalents |
| 3 | pop2019 | population of state, 2019 |
| 4 | landarea | land area of state (square kilometers) |
| 5 | waterarea | water area of state (square kilometers) |

**swedishlcs.csv**   Data on call attempts from the Swedish Survey of Living Conditions. Source: Lundquist and Särndal (2013).

| Column | Name | Value |
|---|---|---|
| 1 | attempt | call attempt number |
| 2 | resprate | response rate at call attempt (percent) |
| 3 | benefits | relative bias for variable *benefits* |
| 4 | income | relative bias for variable *income* |
| 5 | employed | relative bias for variable *employed* |
| 6 | note | Character variable, length 25: notes about data collection |

The variable *attempt* takes on values 1–25 for the initial fieldwork period. Values 31–40 denote the follow-up period, and value 45 gives the final estimates. The gaps in the attempt variable allow one to see the separation of the periods on the graph.

**syc.csv**   Selected variables from the Survey of Youth in Custody (Beck et al., 1988). Source: U.S. Department of Justice (1989). Strata 6–16 each contain one facility; the psus in those strata are residents. In strata 1–5, the psus are facilities. The number of facilities in the population ($N_h$) for those five strata are: $N_1 = 99$, $N_2 = 39$, $N_3 = 30$, $N_4 = 13$, $N_5 = 14$. Eleven facilities are sampled from stratum 1, and seven facilities are sampled from each of strata 2–5.

The table gives missing value codes for individual variables in the .csv file (these codes are the same as in the original data source, but have been changed to the appropriate missing value codes for the respective software packages in the SAS and R data files).

| Column | Name | Value |
|---|---|---|
| 1 | stratum | stratum number |
| 2 | psu | psu number, = facility number for residents in strata 1–5 and person number for residents in strata 6–16 |
| 3 | facility | facility number |
| 4 | facsize | number of eligible residents in psu |
| 5 | finalwt | final weight |
| 6 | randgrp | random group number |
| 7 | age | age of resident (99=missing) |

| syc.csv (continued) | | |
| --- | --- | --- |
| Column | Name | Value |
| 8 | race | race of resident |
| | | 1 = white; 2 = Black; 3 = Asian/Pacific Islander; 4 = American Indian, Alaska Native; 5 = Other; 9 = Missing |
| 9 | ethnicty | 1 = Hispanic, 2 = not Hispanic, 9=missing |
| 10 | educ | highest grade attended before sent to correctional institution |
| | | 0 = Never attended school; 1–12 = highest grade attended; 13 = GED; 14 = Other |
| 11 | gender | 1 = male, 2 = female |
| 12 | livewith | Who did you live with most of the time you were growing up? |
| | | 1 = Mother only, 2 = Father only 3 = Both mother and father, 4 = Grandparents, 5 = Other relatives, 6 = Friends, 7 = Foster home, 8 = Agency or institution, 9 = Someone else, 99 = Blank |
| 13 | famtime | Has anyone in your family, such as your mother, father, brother, sister, ever served time in jail or prison? |
| | | 1 = Yes, 2 = No, 7 = Don't know, 9 = Blank |
| 14 | crimtype | most serious crime in current offense |
| | | 1 = violent (e.g., murder, rape, robbery, assault) |
| | | 2 = property (e.g., burglary, larceny, arson, fraud, motor vehicle theft) |
| | | 3 = drug (drug possession or trafficking) |
| | | 4 = public order (weapons violation, perjury, failure to appear in court) |
| | | 5 = juvenile status offense (truancy, running away, incorrigible behavior) |
| | | 9 = missing |
| 15 | everviol | ever put on probation or sent to correctional inst for violent offense: 1 = yes, 0 = no |
| 16 | numarr | number of times arrested (99=missing) |
| 17 | probtn | number of times on probation (99=missing) |
| 18 | corrinst | number of times previously committed to correctional institution (99=missing) |
| 19 | evertime | Prior to being sent here did you ever serve time in a correctional institution? |
| | | 1 = yes, 2 = no, 9 = missing |
| 20 | prviol | =1 if previously arrested for violent offense, 0 otherwise |
| 21 | prprop | =1 if previously arrested for property offense, 0 otherwise |
| 22 | prdrug | =1 if previously arrested for drug offense, 0 otherwise |
| 23 | prpub | =1 if previously arrested for public order offense, 0 otherwise |
| 24 | prjuv | =1 if previously arrested for juvenile status offense, 0 otherwise |
| 25 | agefirst | age first arrested (99=missing) |
| 26 | usewepn | Did you use a weapon ... for this incident? |
| | | 1 = Yes, 2 = No, 9 = Blank |
| 27 | alcuse | Did you drink alcohol at all during the year before being sent here this time? |
| | | 1 = Yes; 2 = No, didn't drink during year before; 3 = No, don't drink at all, 9 = missing |
| 28 | everdrug | Ever used illegal drugs; 0=no, 1=yes, 9=missing |

**teachers.csv**   Selected variables from a study on elementary school teacher workload in Maricopa County, Arizona. Data courtesy of Rita Gnap (Gnap, 1995). The psu sizes are given in file `teachmi.csv`. The large stratum had 245 schools; the small/medium stratum had 66 schools. Missing values are coded as −9.

| Column | Name | Value |
|---|---|---|
| 1 | dist | school district size. Character variable: large or med/small |
| 2 | school | school identifier |
| 3 | hrwork | number of hours required to work at school per week |
| 4 | size | class size |
| 5 | preprmin | minutes spent per week in school on preparation |
| 6 | assist | minutes per week that a teacher's aide works with the teacher in the classroom |

**teachmi.csv**   Cluster sizes for data in `teachers.csv`.

| Column | Name | Value |
|---|---|---|
| 1 | dist | School district size: large or med/small |
| 2 | school | school identifier |
| 3 | popteach | number of teachers in that school |
| 4 | ssteach | number of surveys returned from that school |

**teachnr.csv**   Data from a follow-up study of nonrespondents from Gnap (1995).

| Column | Name | Value |
|---|---|---|
| 1 | hrwork | number of hours required to work at school per week |
| 2 | size | class size |
| 3 | preprmin | minutes spent per week in school on preparation |
| 4 | assist | minutes per week that a teacher's aide works with the teacher in the classroom |

**uneqvar.csv**   Artificial data used in exercises of Chapter 11.

| Column | Name | Value |
|---|---|---|
| 1 | x | x |
| 2 | y | y |

**vietnam.csv**   Vietnam-service data from Stockford and Page (1984).

| Column | Name | Value |
|---|---|---|
| 1 | apc | APC stratum. Character variable with options "Yes," "No," "NotAvail" |
| 2 | p2sample | Indicator variable for phase II sample, = 1 if in phase II sample, 0 otherwise |
| 3 | vietnam | = 1 if service in Vietnam, = 0 if service not in Vietnam, = −9 if not in phase II sample |

| Column | Name | Value |
|---|---|---|
| **vietnam.csv (continued)** | | |
| Column | Name | Value |
| 4 | phase1wt | weight for phase I sample |
| 5 | phase2wt | conditional weight for phase II sample, calculated as (phase I sample size in stratum) / (phase II sample size in stratum). phase2wt $= -9$ for observations not in phase 2 sample. |
| 6 | finalwt | final weight for phase II sample, calculated as phase1wt*phase2wt ($= -9$ for observations not in phase II sample) |
| 7 | p1apcsize | number of observations in the observation's APC stratum that are in the phase I sample ($n_h$) |
| 8 | p2apcsize | number of observations in the observation's APC stratum that are in the phase II sample ($m_h$) |

**vius.csv** Selected variables from the 2002 U.S. Vehicle Inventory and Use Survey (VIUS). Source: U.S. Census Bureau (2006). The data were downloaded from www.census.gov/svsd/www/vius in May, 2006. The website from which the data were downloaded no longer exists, and online information about VIUS may now be found at https://www.bts.gov/vius, which provides a link to the archived 2002 data. The missing value of *state* for records with *adm_ state* $- 42$ was recoded to "PA," the state that has code 42. This data set has 98,682 records, which may be too large for some software packages to handle; the file viusca.csv is a smaller data set, with the same columns described below, containing only vehicles from California. The variable descriptions below are taken from the VIUS Data Dictionary.

Missing values are coded as $-99$. For some variables, the value is missing because the question is not applicable or the vehicle is not in use; see the individual variable descriptions.

Note that a new VIUS is planned for 2022, with data to be released in 2023; see https://www.bts.gov/vius.

| Column | Name | Value |
|---|---|---|
| 1 | stratum | stratum number (contains all 255 strata) |
| 2 | adm_state | state number |
| 3 | state | state name |
| 4 | trucktype | type of truck, used in stratification<br>1. pickups<br>2. minivans, other light vans, and sport utility vehicles<br>3. light single-unit trucks with gross vehicle weight less than 26,000 pounds<br>4. heavy single-unit trucks with gross vehicle weight greater than or equal to 26,000 pounds<br>5. truck-tractors |
| 5 | tabtrucks | column of sampling weights |
| 6 | bodytype | body type of vehicle<br>01. Pickup<br>02. Minivan<br>03. Light van other than minivan<br>04. Sport utility |

| vius.csv (continued) | | |
| --- | --- | --- |
| Column | Name | Value |
| | | 05. Armored |
| | | 06. Beverage |
| | | 07. Concrete mixer |
| | | 08. Concrete pumper |
| | | 09. Crane |
| | | 10. Curtainside |
| | | 11. Dump |
| | | 12. Flatbed, stake, platform, etc. |
| | | 13. Low boy |
| | | 14. Pole, logging, pulpwood, or pipe |
| | | 15. Service, utility |
| | | 16. Service, other |
| | | 17. Street sweeper |
| | | 18. Tank, dry bulk |
| | | 19. Tank, liquids or gases |
| | | 20. Tow/Wrecker |
| | | 21. Trash, garbage, or recycling |
| | | 22. Vacuum |
| | | 23. Van, basic enclosed |
| | | 24. Van, insulated non-refrigerated |
| | | 25. Van, insulated refrigerated |
| | | 26. Van, open top |
| | | 27. Van, step, walk-in, or multistop |
| | | 28. Van, other |
| | | 99. Other not elsewhere classified |
| 7 | adm_modelyear | model year |
| | | 01. 2003, 2002 |
| | | 02. 2001 |
| | | 03. 2000 |
| | | 04. 1999 |
| | | 05. 1998 |
| | | 06. 1997 |
| | | 07. 1996 |
| | | 08. 1995 |
| | | 09. 1994 |
| | | 10. 1993 |
| | | 11. 1992 |
| | | 12. 1991 |
| | | 13. 1990 |
| | | 14. 1989 |
| | | 15. 1988 |
| | | 16. 1987 |
| | | 17. Pre-1987 |
| 8 | vius_gvw | Gross vehicle weight based on average reported weight |
| | | 01. Less than 6,001 lbs. |
| | | 02. 6,001 to 8,500 lbs. |
| | | 03. 8,501 to 10,000 lbs. |
| | | 04. 10,001 to 14,000 lbs. |

| vius.csv (continued) | | |
| --- | --- | --- |
| Column | Name | Value |
| | | 05. 14,001 to 16,000 lbs. |
| | | 06. 16,001 to 19,500 lbs. |
| | | 07. 19,501 to 26,000 lbs. |
| | | 08. 26,001 to 33,000 lbs. |
| | | 09. 33,001 to 40,000 lbs. |
| | | 10. 40,001 to 50,000 lbs. |
| | | 11. 50,001 to 60,000 lbs. |
| | | 12. 60,001 to 80,000 lbs. |
| | | 13. 80,001 to 100,000 lbs. |
| | | 14. 100,001 to 130,000 lbs. |
| | | 15. 130,001 lbs. or more |
| 9 | miles_annl | Number of Miles Driven During 2002 |
| 10 | miles_life | Number of Miles Driven Since Manufactured |
| 11 | mpg | Miles Per Gallon averaged during 2002. Range from 0.3 to 35. −99 denotes not reported or not applicable. |
| 12 | opclass | Operator Classification With Highest Percent<br>1. Private<br>2. Motor carrier<br>3. Owner operator<br>4. Rental<br>5. Personal transportation<br>6. Not applicable (Vehicle not in use) |
| 13 | opclass_mtr | Percent of Miles Driven as a Motor Carrier. −99 denotes vehicle not in use |
| 14 | opclass_own | Percent of Miles Driven as an Owner Operator. −99 denotes vehicle not in use |
| 15 | opclass_psl | Percent of Miles Driven for Personal Transportation. −99 denotes vehicle not in use |
| 16 | opclass_pvt | Percent of Miles Driven as Private (Carry Own Goods or Internal Company Business Only). −99 denotes vehicle not in use |
| 17 | opclass_rnt | Percent of Miles Driven as Rental. −99 denotes vehicle not in use |
| 18 | transmssn | Type of Transmission<br>1. Automatic<br>2. Manual<br>3. Semi-Automated Manual<br>4. Automated Manual |
| 19 | trip_primary | Primary Range of Operation<br>1. Off-the-road<br>2. Less than 50 miles<br>3. 51 to 100 miles<br>4. 101 to 200 miles<br>5. 201 to 500 miles<br>6. 501 miles or more<br>7. Not reported<br>8. Not applicable (Vehicle not in use) |
| 20 | trip0_50 | Percent of Annual Miles Accounted for with Trips |

| vius.csv (continued) | | |
|---|---|---|
| Column | Name | Value |

|  |  | 50 Miles or Less from the Home Base |
|---|---|---|
| 21 | trip051_100 | Percent of Annual Miles Accounted for with Trips 51 to 100 Miles from the Home Base |
| 22 | trip101_200 | Percent of Annual Miles Accounted for with Trips 101 to 200 Miles from the Home Base |
| 23 | trip201_500 | Percent of Annual Miles Accounted for with Trips 201 to 500 Miles from the Home Base |
| 24 | trip500more | Percent of Annual Miles Accounted for with Trips 501 or More Miles from Home Base |
| 25 | adm_make | Make of vehicle |
|  |  | 01. Chevrolet |
|  |  | 02. Chrysler |
|  |  | 03. Dodge |
|  |  | 04. Ford |
|  |  | 05. Freightliner |
|  |  | 06. GMC |
|  |  | 07. Honda |
|  |  | 08. International |
|  |  | 09. Isuzu |
|  |  | 10. Jeep |
|  |  | 11. Kenworth |
|  |  | 12. Mack |
|  |  | 13. Mazda |
|  |  | 14. Mitsubishi |
|  |  | 15. Nissan |
|  |  | 16. Peterbilt |
|  |  | 17. Plymouth |
|  |  | 18. Toyota |
|  |  | 19. Volvo |
|  |  | 20. White |
|  |  | 21. Western Star |
|  |  | 22. White GMC |
|  |  | 23. Other (domestic) |
|  |  | 24. Other (foreign) |
| 26 | business | Business in which vehicle was most often used during 2002 |
|  |  | 01. For-hire transportation or warehousing |
|  |  | 02. Vehicle leasing or rental |
|  |  | 03. Agriculture, forestry, fishing, or hunting |
|  |  | 04. Mining |
|  |  | 05. Utilities |
|  |  | 06. Construction |
|  |  | 07. Manufacturing |
|  |  | 08. Wholesale trade |
|  |  | 09. Retail trade |
|  |  | 10. Information services |
|  |  | 11. Waste management, landscaping, or administrative/support services |

| vius.csv (continued) | | |
|---|---|---|
| Column | Name | Value |
| | | 12. Arts, entertainment, or recreation services |
| | | 13. Accommodation or food services |
| | | 14. Other services |
| | | −99. Not reported or not applicable |

**winter.csv**  Selected variables from the Arizona State University Winter Closure Survey, taken in January 1995 (provided courtesy of the ASU Office of University Evaluation). This survey was taken to investigate the attitudes and opinions of university employees towards the closing of the university (for budgetary reasons) between December 25 and January 1. For the yes/no questions, the responses are coded as 1 = No, 2 = Yes. The variables *treatsta* and *treatme* are coded as 1=strongly agree, 2=agree, 3=undecided, 4=disagree, 5=strongly disagree. The variables *process* and *satbreak* are coded as 1=very satisfied, 2=satisfied, 3=undecided, 4=dissatisfied, 5=very dissatisfied. Variables *ownsupp* through *offclose* are coded 1 if the person checked that the statement applied to him/her, and 2 if the statement was not checked.

Missing values are coded as 9.

| Column | Name | Value |
|---|---|---|
| 1 | class | stratum number |
| | | 1 = faculty ; 2 = classified staff; 3 = administrative staff; 4 = academic professional |
| 2 | yearasu | number of years worked at ASU |
| | | 1 = 1–2 years; 2 = 3–4 years; 3 = 5–9 years; 4 = 10–14 years; 5 = 15 or more years |
| 3 | vacation | In the past, have you *usually* taken vacation days the entire period between December 25 and January 1? |
| 4 | work | Did you work on campus during Winter Break Closure? |
| 5 | havediff | Did the Winter Break Closure cause you any difficulty/concerns? |
| 6 | negaeffe | Did the Winter Break Closure *negatively* affect your work productivity? |
| 7 | ownsupp | I was unable to obtain staff support in my department/office |
| 8 | othersup | I was unable to obtain staff support in other departments/offices |
| 9 | utility | I was unable to access computers, copy machine, etc. in my department/office |
| 10 | environ | I was unable to endure environmental conditions, e.g., not properly climatized |
| 11 | uniserve | I was unable to access university services necessary to my work |
| 12 | workelse | I was unable to work on my assignments because I work in another department/office |
| 13 | offclose | I was unable to work on my assignments because my office was closed |
| 14 | treatsta | Compared to other departments/offices, I feel staff in my department/office were treated fairly |

| winter.csv (continued) | | |
|---|---|---|
| **Column** | **Name** | **Value** |
| 15 | treatme | Compared to other people working in my department/office, I feel I was treated fairly |
| 16 | process | How satisfied are you with the process used to inform staff about Winter Break Closure? |
| 17 | satbreak | How satisfied are you with the fact that ASU had a Winter Break Closure this year? |
| 18 | breakaga | Would you want to have Winter Break Closure again? |

**wtshare.csv** Hypothetical sample of size 100, with indirect sampling. The data set has multiple records for adults with more than one child; if adult 254 has 3 children, adult 254 is listed 3 times in the data set. Note that to obtain $L_k$, you need to take *numadult* $+1$.

| **Column** | **Name** | **Value** |
|---|---|---|
| 1 | id | Identification number of adult in sample |
| 2 | child | $= 1$ if record is for a child, 0 if adult has no children |
| 3 | preschool | $= 1$ if child is in preschool, 0 otherwise |
| 4 | numadult | number of other adults in population who link to that child |

# Bibliography

Allison, P. D. (2012). *Logistic Regression using SAS®: Theory and Application*. Cary, NC: SAS Institute, Inc.

Arnold, T. W. (1991). Intraclutch variation in egg size of American coots. *The Condor 93*, 19–27.

Asian Development Bank (2020). *Introduction to Small Area Estimation Techniques: A Practical Guide for National Statistics Offices*. Manila: Asian Development Bank.

Azur, M. J., E. A. Stuart, C. Frangakis, and P. J. Leaf (2011). Multiple imputation by chained equations: What is it and how does it work? *International Journal of Methods in Psychiatric Research 20*(1), 40–49.

Baillargeon, S. and L.-P. Rivest (2007). Rcapture: Loglinear models for capture-recapture in R. *Journal of Statistical Software 19*(5), 1–31.

Baillargeon, S. and L.-P. Rivest (2011). The construction of stratified designs in R with the package stratification. *Survey Methodology 37*(1), 53–65.

Barcaroli, G. (2014). SamplingStrata: An R package for the optimization of stratified sampling. *Journal of Statistical Software 61*(4), 1–24.

Barcaroli, G., M. Ballin, H. Odendaal, D. Pagliuca, E. Willighagen, and D. Zardetto (2020). *SamplingStrata: Optimal Stratification of Sampling Frames for Multipurpose Sampling Surveys*. R package version 1.5-1, `https://CRAN.R-project.org/package=SamplingStrata` (accessed March 11, 2021).

Bart, J. and S. Earnst (2002). Double-sampling to estimate density and population trends in birds. *The Auk 119*, 36–45.

Bates, D., M. Mächler, B. Bolker, and S. Walker (2015). Fitting linear mixed-effects models using lme4. *Journal of Statistical Software 67*(1), 1–48.

Bates, D., M. Mächler, B. Bolker, S. Walker, R. H. B. Christensen, H. Singmann, B. Dai, F. Scheipl, G. Grothendieck, P. Green, J. Fox, A. Bauer, and P. N. Krivitsky (2020). *lme4: Linear Mixed-Effects Models using 'Eigen' and S4*. R package version 1.1-26, `https://CRAN.R-project.org/package=lme4` (accessed March 20, 2021).

Beck, A. J., S. A. Kline, and L. A. Greenfeld (1988). Survey of Youth in Custody. Technical Report NCJ-113365, Bureau of Justice Statistics, Washington, DC.

Bretz, F., T. Hothorn, and P. Westfall (2016). *Multiple Comparisons using R*. Boca Raton, FL: CRC Press.

Brewer, K. R. W. (1963). Ratio estimation and finite populations: Some results deducible from the assumption of an underlying stochastic process. *The Australian Journal of Statistics 5*(3), 93–105.

Brewer, K. R. W. (1975). A simple procedure for sampling πpswor. *The Australian Journal of Statistics 17*, 166–172.

Brick, J. M., D. Morganstein, and R. Valliant (2000). *Analysis of Complex Sample Data Using Replication*. Rockville, MD: Westat.

Bueno, E. (2020). *optimStrat: Choosing the Sample Strategy*. R package version 2.3. `https://CRAN.R-project.org/package=optimStrat` (accessed March 12, 2021).

Canty, A. J. and A. C. Davison (1999). Resampling-based variance estimation for labour force surveys. *The Statistician 48*, 379–391.

Centers for Disease Control and Prevention (2017). NHANES Questionnaires, Datasets, and Related Documentation. `https://wwwn.cdc.gov/nchs/nhanes/` (accessed August 15, 2020).

Chao, A., K. H. Ma, T. C. Hsieh, and C.-H. Chiu (2016). *SpadeR: Species-Richness Prediction and Diversity Estimation with R*. R package version 0.1.1, `https://CRAN.R-project.org/package=SpadeR` (accessed March 15, 2021).

Chapman, D. G. (1951). Some properties of the hypergeometric distribution with applications to zoological sample censuses. *University of California Publications in Statistics 1*, 131–160.

Chauvet, G. and Y. Tillé (2006). A fast algorithm for balanced sampling. *Computational Statistics 21*(1), 53–62.

Cho, I., J.-K. Kim, J. Im, and Y. Yang (2020). *FHDI: Fractional Hot Deck and Fully Efficient Fractional Imputation*. R package version 1.4.1, `https://CRAN.R-project.org/package=FHDI` (accessed April 15, 2021).

Cormack, R. M. (1992). Interval estimation for mark-recapture studies of closed populations. *Biometrics 48*, 567–576.

Dippo, C. S., R. E. Fay, and D. H. Morganstein (1984). Computing variances from complex samples with replicate weights. In *Proceedings of the Survey Research Methods Section*, pp. 489–494. Alexandria, VA: American Statistical Association.

Domingo-Salvany, A., R. L. Hartnoll, A. Maquire, J. M. Suelves, and J. M. Anto (1995). Use of capture-recapture to estimate the prevalence of opiate addiction in Barcelona, Spain, 1989. *American Journal of Epidemiology 141*, 567–574.

Fienberg, S. E. and A. Rinaldo (2007). Three centuries of categorical data analysis: Log-linear models and maximum likelihood estimation. *Journal of Statistical Planning and Inference 137*(11), 3430–3445.

Forman, S. L. (2004). Baseball-reference.com—Major league statistics and information. `www.baseball-reference.com` (accessed November 2004).

Gambino, J. G. (2021). *pps: PPS Sampling*. R package version 1.0. `https://CRAN.R-project.org/package=pps` (accessed March 12, 2021).

Gini, C. and L. Galvani (1929). Di una applicazione del metodo rappresentativo all'ultimo censimento italiano della popolazione. *Annali di Statistica 6*(4), 1–105.

Gnap, R. (1995). *Teacher Load in Arizona Elementary School Districts in Maricopa County*. Ph.D. dissertation. Tempe, AZ: Arizona State University.

Goga, C. (2018). Brief overview of survey sampling techniques with R. *Romanian Statistical Review 2018*(1), 83–94.

Grafström, A. and J. Lisic (2019). *BalancedSampling: Balanced and Spatially Balanced Sampling*. R package version 1.5.5, `https://CRAN.R-project.org/package=BalancedSampling` (accessed March 12, 2021).

Hand, D. J., F. Daly, A. D. Lunn, K. J. McConway, and E. Ostrowski (1994). *A Handbook of Small Data Sets*. London: Chapman and Hall.

Hanurav, T. V. (1967). Optimum utilization of auxiliary information: πps sampling of two units from a stratum. *Journal of the Royal Statistical Society, Series B 29*, 374–391.

Harmening, S., A.-K. Kreutzmann, S. Pannier, N. Rojas-Perilla, N. Salvati, T. Schmid, M. Templ, N. Tzavidis, and N. Würz (2021). *emdi: Estimating and Mapping Disaggregated Indicators*. R package version 2.0.2, `https://CRAN.R-project.org/package=emdi` (accessed April 30, 2021).

Harrell, F. E. (2021). *Hmisc: Harrell miscellaneous*. R package version 4.5-0, `https://CRAN.R-project.org/package=Hmisc` (accessed April 7, 2021).

Hartley, H. O. and J. N. K. Rao (1962). Sampling with unequal probabilities and without replacement. *The Annals of Mathematical Statistics 33*, 350–374.

Hayat, M. and T. Knapp (2017). Randomness and inference in medical and public health research. *Journal of the Georgia Public Health Association 7*(1), 7–11.

Haziza, D. (2009). Imputation and inference in the presence of missing data. In D. Pfeffermann and C. R. Rao (Eds.), *Sample Surveys: Design, Methods, and Applications. Handbook of Statistics, Volume 29A*, pp. 215–246. Amsterdam: North-Holland.

Heck, P. R., D. J. Simons, and C. F. Chabris (2018). 65% of Americans believe they are above average in intelligence: Results of two nationally representative surveys. *PloS One 13*(7), 1–11.

Hidiroglou, M. A., J.-F. Beaumont, and W. Yung (2019). Development of a small area estimation system at Statistics Canada. *Survey Methodology 45*(1), 101–126.

Horton, N. J. and K. Kleinman (2015). *Using R and RStudio for Data Management, Statistical Analysis, and Graphics, 2nd ed*. Boca Raton, FL: CRC Press.

Horton, N. J., R. Pruim, and D. T. Kaplan (2018). *A Student's Guide to R*. Amherst, MA: Project MOSAIC.

Hyndman, R. J. and Y. Fan (1996). Sample quantiles in statistical packages. *The American Statistician 50*(4), 361–365.

Im, J., I. H. Cho, and J.-K. Kim (2018). FHDI: An R package for fractional hot deck imputation. *R Journal 10*(1), 140–154.

Ismay, C. and P. C. Kennedy (2019). *Getting Used to R, RStudio, and R Markdown*. `https://rbasics.netlify.app/` (accessed March 2, 2021).

Jacoby, J. and A. H. Handlin (1991). Non-probability sampling designs for litigation surveys. *Trademark Reporter 81*, 169–179.

Judkins, D. (1990). Fay's method for variance estimation. *Journal of Official Statistics 6*, 223–240.

Kabacoff, R. I. (2021). *R in Action, 3rd ed.* Shelter Island, NY: Manning Publications.

Kim, J. K. (2011). Parametric fractional imputation for missing data analysis. *Biometrika 98*(1), 119–132.

Kim, J. K. and W. Fuller (2004). Fractional hot deck. imputation. *Biometrika 91*(3), 559–578.

Koch, G. G., D. H. Freeman, and J. L. Freeman (1975). Strategies in the multivariate analysis of data from complex surveys. *International Statistical Review 43*, 59–78.

Koenker, R. (2005). *Quantile Regression.* Cambridge: Cambridge University Press.

Koenker, R., S. Portnoy, P. T. Ng, B. Melly, A. Zeileis, P. Grosjean, C. Moler, Y. Saad, V. Chernozhukov, I. Fernandez-Val, and B. D. Ripley (2021). *quantreg: Quantile Regression.* R package version 2.23-18, `https://CRAN.R-project.org/package=quantreg` (accessed April 7, 2021).

Korn, E. L. and B. I. Graubard (1998). Confidence intervals for proportions with small expected number of positive counts estimated from survey data. *Survey Methodology 24*, 193–201.

Kott, P. S. (2012). Why one should incorporate the design weights when adjusting for unit nonresponse using response homogeneity groups. *Survey Methodology 38*(1), 95–99.

Kowarik, A. and M. Templ (2016). Imputation with the R package VIM. *Journal of Statistical Software 74*(7), 1–16.

Kreutzmann, A.-K., S. Pannier, N. Rojas-Perilla, T. Schmid, M. Templ, and N. Tzavidis (2019). The R package emdi for estimating and mapping regionally disaggregated indicators. *Journal of Statistical Software 91*, 1–33.

Kruuk, H., A. Moorhouse, J. W. H. Conroy, L. Durbin, and S. Frears (1989). An estimate of numbers and habitat preferences of otters *Lutra lutra* in Shetland, UK. *Biological Conservation 49*, 241–254.

Little, R. J. and S. Vartivarian (2003). On weighting the rates in non-response weights. *Statistics in Medicine 22*(9), 1589–1599.

Lohr, S. L. (2022). *SAS® Software Companion for* Sampling: Design and Analysis, 3rd ed. Boca Raton, FL: CRC Press.

Lopez-Vizcaino, E., M. Lombardia, and D. Morales (2019). *mme: Multinomial Mixed Effects Models.* R package version 0.1-6. `https://CRAN.R-project.org/package=mme` (accessed January 16, 2021).

Lu, Y. and S. L. Lohr (2021). *SDAResources: Datasets and Functions for "Sampling: Design and Analysis."* R package version 0.1.0. `https://CRAN.R-project.org/package=SDAResources` (accessed May 17, 2021).

Lumley, T. (2004). Analysis of complex survey samples. *Journal of Statistical Software 9*(1), 1–19.

Lumley, T. (2010). *Complex Surveys: A Guide to Analysis using R.* Hoboken, NJ: Wiley.

Lumley, T. (2020). *survey: Analysis of Complex Survey Samples.* R package version 4.0. `https://CRAN.R-project.org/package=survey` (accessed September 20, 2020).

Lundquist, P. and C.-E. Särndal (2013). Aspects of responsive design with applications to the Swedish Living Conditions Survey. *Journal of Official Statistics 29*(4), 557–582.

Luraschi, J. (2021). Importing data with RStudio. `https://support.rstudio.com/hc/en-us/articles/218611977-Importing-Data-with-RStudio` (accessed March 12, 2021).

Lydersen, C. and M. Ryg (1991). Evaluating breeding habitat and populations of ringed seals *Phoca hispida* in Svalbard fjords. *Polar Record 27*, 223–228.

Macdonell, W. R. (1901). On criminal anthropometry and the identification of criminals. *Biometrika 1*, 177–227.

Molina, I. and Y. Marhuenda (2015). sae: An R package for small area estimation. *The R Journal 7*(1), 1–98.

Molina, I. and Y. Marhuenda (2020). *sae: Small Area Estimation.* R package version 1.3. `https://CRAN.R-project.org/package=sae` (accessed December 20, 2020).

National Center for Health Statistics (1987). *Vital Statistics of the United States, Volume 3: Marriage and Divorce.* Washington, DC: U.S. Government Printing Office.

Nolan, D. and T. Speed (2000). *Stat Labs: Mathematical Statistics Through Applications.* New York: Springer.

Oetiker, T., H. Partl, I. Hyna, and E. Schlegl (2021). *The Not So Short Introduction to LaTeX $2_\varepsilon$, Version 6.4.* Olten, Switzerland: Tobias Oetiker. `https://tobi.oetiker.ch/lshort/lshort.pdf` (accessed March 18, 2021).

Peart, D. (1994). *Impacts of Feral Pig Activity on Vegetation Patterns Associated with Quercus agrifolia on Santa Cruz Island, California.* Ph.D. dissertation. Tempe, AZ: Arizona State University.

Pinheiro, J., D. Bates, S. DebRoy, D. Sarkar, and R Core Team (2021). *nlme: Linear and Nonlinear Mixed Effects Models.* R package version 3.1-152, `https://CRAN.R-project.org/package=nlme` (accessed March 20, 2021).

Pratesi, M. (Ed.) (2016). *Analysis of Poverty Data by Small Area Estimation.* Hoboken, NJ: Wiley.

Preston, J. (2009). Rescaled bootstrap for stratified multistage sampling. *Survey Methodology 35*(2), 227–234.

R Core Team (2021). *R: A Language and Environment for Statistical Computing.* Vienna, Austria: R Foundation for Statistical Computing. Version 4.0.4, `https://www.R-project.org`.

Rao, J. N. K. and I. Molina (2015). *Small Area Estimation, 2nd ed.* Hoboken, NJ: Wiley.

Rao, J. N. K. and A. J. Scott (1981). The analysis of categorical data from complex sample surveys: Chi-squared tests for goodness of fit and independence in two-way tables. *Journal of the American Statistical Association 76*, 221–230.

Rao, J. N. K. and A. J. Scott (1984). On chi-squared tests for multiway contingency tables with cell proportions estimated from survey data. *The Annals of Statistics 12*, 46–60.

Rao, J. N. K., C. F. J. Wu, and K. Yue (1992). Some recent work on resampling methods for complex surveys. *Survey Methodology 18*, 209–217.

Reiter, J. P., T. E. Raghunathan, and S. K. Kinney (2006). The importance of modeling the sampling design in multiple imputation for missing data. *Survey Methodology 32*(2), 143–149.

Rivest, L.-P. and S. Baillargeon (2017). *stratification: Univariate Stratification of Survey Populations*. R package version 2.2-6, `https://CRAN.R-project.org/package=stratification` (accessed March 10, 2021).

Rivest, L.-P. and S. Baillargeon (2019). *Rcapture: Loglinear Models for Capture-Recapture Experiments*. R package version 1.4-3, `https://CRAN.R-project.org/package=Rcapture` (accessed March 15, 2021).

Roberts, R. J., Q. D. Sandifer, M. R. Evans, M. Z. Nolan-Ferrell, and P. M. Davis (1995). Reasons for non-uptake of measles, mumps, and rubella catch up immunisation in a measles epidemic and side effects of the vaccine. *British Medical Journal 310*, 1629–1632.

Ruggles, S., M. Sobek, T. Alexander, C. A. Fitch, R. Goeken, P. K. Hall, M. King, and C. Ronnander (2004). Integrated Public Use Microdata Series: Version 3.0 [machine-readable database]. `www.ipums/org/usa` (accessed September 17, 2008).

Sampford, M. R. (1967). On sampling without replacement with unequal probabilities of selection. *Biometrika 54*, 499–513.

SAS Institute Inc. (2021). *SAS/STAT® User's Guide*. Cary, NC: SAS Institute Inc. `https://documentation.sas.com/` (accessed April 27, 2021).

Stockford, D. D. and W. F. Page (1984). Double sampling and the misclassification of Vietnam service. In *Proceedings of the Social Statistics Section*, pp. 261–264. Alexandria, VA: American Statistical Association.

Talbot, N. L. C. (2012). *LaTeX for Complete Novices*. Saxlingham Nethergate, UK: Dickimaw Books. `https://www.dickimaw-books.com/latex/novices/` (accessed March 10, 2021).

Templ, M., A. Kowarik, A. Alfons, G. de Cillia, B. Prantner, and W. Rannetbauer (2021). *VIM: Visualization and Imputation of Missing Values*. R package version 6.1.0, `https://CRAN.R-project.org/package=VIM` (accessed April 21, 2021).

Thomas, D. R. and J. N. K. Rao (1987). Small-sample comparisons of level and power for simple goodness-of-fit statistics under cluster sampling. *Journal of the American Statistical Association 82*, 630–636.

Tillé, Y. (2006). *Sampling Algorithms*. New York: Springer.

Tillé, Y. and A. Matei (2010). Teaching survey sampling with the 'sampling' R package. In *Proceedings of the 8th International Conference on Teaching Statistics (ICOTS)*, pp. 1–6. Auckland, NZ: International Association of Statistical Education.

Tillé, Y. and A. Matei (2021). *sampling: Survey Sampling*. R package version 2.9, `https://CRAN.R-project.org/package=sampling` (accessed March 12, 2021).

Tillé, Y. and M. Wilhelm (2017). Probability sampling designs: Principles for choice of design and balancing. *Statistical Science 32*(2), 176–189.

Tzavidis, N., L.-C. Zhang, A. Luna, T. Schmid, and N. Rojas-Perilla (2018). From start to finish: A framework for the production of small area official statistics. *Journal of the Royal Statistical Society: Series A 181*(4), 927–979.

U.S. Bureau of the Census (1921). *Fourteenth Census of the United States Taken in the Year 1920*. Washington, DC: U.S. Government Printing Office. `https://www.census.gov/library/publications/1921/dec/vol-01-population.html` (accessed August 4, 2020).

U.S. Bureau of the Census (1995). *1992 Census of Agriculture, Volume 1: Geographic Area Series*. Washington, DC: U.S. Bureau of the Census.

U.S. Census Bureau (1994). *County and City Data Book: 1994*. Washington, DC: U.S. Census Bureau.

U.S. Census Bureau (2006). *Vehicle Inventory and Use Survey—Methods*. Washington, DC: U.S. Census Bureau.

U.S. Census Bureau (2012). *United States Summary, 2010*. Washington, DC: U.S. Census Bureau. `https://www.census.gov/prod/cen2010/cph-2-1.pdf` (accessed October 3, 2020).

U.S. Census Bureau (2019). State population totals: 2010-2019. Table 1. Annual estimates of the resident population for the United States, regions, states, and Puerto Rico: April 1, 2010 to July 1, 2019 (NST-EST2019-01). `https://www.census.gov/data/datasets/time-series/demo/popest/2010s-state-total.html` (accessed August 3, 2020).

U.S. Department of Education (2020). College scorecard data. `https://collegescorecard.ed.gov/data/` (accessed August 25, 2020).

U.S. Department of Justice (1989). *Survey of Youth in Custody, 1987, United States computer file, Conducted by Department of Commerce, Bureau of the Census, 2nd ICPSR ed.* Ann Arbor, MI: Inter-University Consortium for Political and Social Research.

Valliant, R., J. A. Dever, and F. Kreuter (2018). *Practical Tools for Designing and Weighting Survey Samples*. New York: Springer.

Valliant, R., J. A. Dever, and F. Kreuter (2020). *PracTools: Tools for Designing and Weighting Survey Samples*. R package version 1.2.2. `https://CRAN.R-project.org/package=PracTools` (accessed March 12, 2021).

Valliant, R. and K. F. Rust (2010). Degrees of freedom approximations and rules-of-thumb. *Journal of Official Statistics 26*(4), 585–602.

van Buuren, S. (2018). *Flexible Imputation of Missing Data*. Boca Raton, FL: CRC Press.

van Buuren, S., K. Groothuis-Oudshoorn, G. Vink, R. Schouten, A. Robitzsch, P. Rockenschaub, L. Doove, S. Jolani, M. Moreno-Betancur, I. White, P. Gaffert, F. Meinfelder, B. Gray, and V. Arel-Bundock (2021). *mice: Multivariate Imputation by Chained Equations*. R package version 3.13.0, `https://CRAN.R-project.org/package=mice` (accessed April 15, 2021).

Vijayan, K. (1968). An exact πps sampling scheme: Generalization of a method of Hanurav. *Journal of the Royal Statistical Society, Series B 30*, 556–566.

Virginia Polytechnic and State University/Responsive Management (2006). *An Assessment of Public and Hunter Opinions and the Costs and Benefits to North Carolina of Sunday Hunting.* Blacksburg, VA: Virginia Polytechnic and State University.

Wand, M., C. Moler, and B. Ripley (2020). *KernSmooth: Functions for Kernel Smoothing Supporting Wand & Jones (1995).* R package version 2.23-18, `https://CRAN.R-project.org/package=KernSmooth` (accessed April 7, 2021).

Wand, M. P. and M. C. Jones (1995). *Kernel Smoothing.* London: Chapman & Hall.

Wang, J. (2021). The pseudo maximum likelihood estimator for quantiles of survey variables. *Journal of Survey Statistics and Methodology 9*(1), 185–201.

Wickham, H. (2015). *R Packages: Organize, Test, Document, and Share Your Code.* Sebastopol, CA: O'Reilly Media.

Wickham, H. (2019). *Advanced R, 2nd ed.* Boca Raton, FL: CRC Press.

Wickham, H., W. Chang, L. Henry, T. L. Pedersen, K. Takahashi, C. Wilke, K. Woo, H. Yutani, and D. Dunnington (2020). *ggplot2: Create Elegant Data Visualisations Using the Grammar of Graphics.* R package version 3.3.3, `https://CRAN.R-project.org/package=ggplot2` (accessed March 30, 2021).

Wikibooks contributors (2021). *LaTeX.* Wikibooks, The Free Textbook Project. `https://en.wikibooks.org/wiki/LaTeX` (accessed March 18, 2021).

Wilk, S. J., W. W. Morse, D. E. Ralph, and T. R. Azarovitz (1977). *Fishes and Associated Environmental Data Collected in New York Bight, June 1974–June 1975.* NOAA Tech. Rep. No. NMFS SSRF-716. Washington, DC: U.S. Government Printing Office.

Woodruff, R. S. (1952). Confidence intervals for medians and other position measures. *Journal of the American Statistical Association 47*, 636–646.

Xie, Y. (2015). *Dynamic Documents with R and knitr, 2nd ed.* Boca Raton, FL: CRC Press.

Yadav, M. L. and B. Roychoudhury (2018). Handling missing values: A study of popular imputation packages in R. *Knowledge-Based Systems 160*, 104–118.

Zhang, C., C. Antoun, H. Y. Yan, and F. G. Conrad (2020). Professional respondents in opt-in online panels: What do we really know? *Social Science Computer Review 38*(6), 703–719.

Zhang, G., F. Christensen, and W. Zheng (2015). Nonparametric regression estimators in complex surveys. *Journal of Statistical Computation and Simulation 85*(5), 1026–1034.

# *Index*